# Current Advances in the Medical Application of Nanotechnology

## Edited By

## Mark Slevin

*Manchester Metropolitan University*
*Manchester*
*United Kingdom*

# eBooks End User License Agreement

# CONTENTS

## CHAPTERS

# Editor's Note

*For medicine, nanotechnology promises new therapies, more rapid and sensitive diagnostic and investigative tools for normal and diseased tissues, and new materials for tissue engineering.* This book will highlight the major current uses, new technologies and future perspectives of nanotechnology in relation to medical application. This e-Book will include sections on nanobiological approaches to imaging, diagnosis and treatment of disease, ranging from the medical use of nanomaterials, to nanoelectronic biosensors, and possible future applications of molecular nanotechnology for cell repair. Also covered will be current problems associated with this technology, including an understanding of issues related to toxicity and the environmental impact of nanoscale materials. This e-Book will be of interest to scientists, biomedical technologists, medical doctors and students of taught courses in nanotechnology at both under-graduate and post-graduate level (MSc).

Targeting therapeutic drugs directly at disease sites increases their effectiveness and reduces side effects. Nanoparticles can encapsulate drugs, specific proteins or antibodies and can even be attached to artificial RNA strands known as aptamers. The nanoparticles can be taken up by specific cells where the nanoparticles dissolve to release the protein or drug. This technology could be used to enhance tissue regeneration and remodelling or conversely, destroy diseased tissue such as tumours. The treatment of disease depends on the identification of a target and delivery of a therapeutic agent which either causes function to be restored, switches off inappropriate activity, or in the case of cancer, destroys the cell. However, many pharmaceuticals are limited in their development or application because of poor solubility, poor stability, or side-effects in inappropriate tissues. Manipulating the composition of a drug formulation at the nanoscale can resolve these issues.

Although at an early stage of development, nanoshells can be used to encapsulate drugs, protecting them from the environment and offering targeted release (essential for toxic anti-cancer therapies). Nanoshells can be made from polymers (which fuse with cell membranes and release their contents within the cell) or a mix of polymer and gold (which can be induced to melt when irradiated with infrared light thus releasing their contents). Other nanomaterials can be used to target or directly treat diseased tissues by physical rather than biochemical means. For example, paramagnetic iron nanoparticles can be made to accumulate in tumour cells through the use of magnetic fields and can be used to destroy tumour cells by heat through the application of alternating magnetic fields (known as magnetic fluid hyperthermia [MFH]; In the future functionalising such nanoparticles with targeting biomolecules could enable them to be delivered systemically.

*Mark Slevin*
Manchester Metropolitan University
Manchester
United Kingdom

# FOREWORD

Nanotechnology is becoming increasingly important in biomedical research and clinical medicine. Similar to other fields, nanotechnology is believed to have a great future in broad aspects of medicine. For example, nanoparticles can potentially improve drug delivery into targeted cells or specific sites within tissue, overcome drug insolubility, enhance drug penetration across barriers and prolong drug bioavailability or achieve sustained drug release. Nanoparticle-mediated drug delivery is being widely studied, as nanoparticles can encapsulate drugs in broad categories including small molecule compounds, peptides, antibodies, siRNA and DNA. Furthermore, most nanoparticles have excellent storage capacity and can be produced with biocompatible and biodegradable materials. Due to their small size, nanoparticles can carry drugs into cells to reach disease sites, which increases local drug concentrations and reduces systemic toxicities. Nanoparticles are also used to achieve sustained drug release after a single injection and thus reduce the injection frequency. This application is especially attractive for drug delivery to sites which drugs cannot reach in an efficient manner. In addition, the engineering of specialized nanoparticles offers a tremendous promise for improving medical imaging and diagnosis techniques.

Nanotechnology in medicine is a fast growing field with numerous publications each year. The purpose of this e-Book is to introduce the rationale for the use of nanotechnology in medicine and summarize recent advances in some specific applications of nanotechnology in medicine. This e-Book also provides a perspective on the future development and potential use of nanotechnology in medicine and discusses the current limitations and hurdles in the medical application of nanotechnology. This e-Book will be of interest to readers in broad fields, including scientists in drug research and development, biomedical technologists, researchers in nanotechnology, medical doctors with interests in improved drug delivery and diagnosis, as well as students in nanotechnology and drug delivery fields.

The editor of this e-Book, Dr. Mark Slevin, is a well established scientist with extensive knowledge in the medical application of nanotechnology. Dr. Slevin has been doing research in nanotechnology for many years, and has published multiple peer-review journal articles, invited reviews and book chapters. He is currently serving as Editor-in-Chief for the Journal of Vascular Cell, and Editor of Brain Pathology, Frontiers in Biosciences, Neural Regeneration Research, the Open Circulation and Microvascular Journal, World Journal of Cardiology. Each chapter of this e-Book contains an in-depth review of a growing field of medical nanotechnology and has been written by leading experts. This e-Book is an excellent resource for anyone interested in the current status and future potential of the utilization of nanotechnology in medicine.

*Jian-Xing Ma*
Department of Medicine Endocrinology
University of Oklahoma Health Sciences Center (OUHSC)
Oklahoma City, OK 73104
USA

# CONTRIBUTORS

| | |
|---|---|
| *Mark Slevin* | School of Biology, Chemistry and Health Science, Manchester Metropolitan University, Manchester, UK; St Pau Hospital, Barcelona, Spain |
| *Hemant Sarin* | National Institute of Biomedical Imaging and Bioengineering, National Institutes of Health, Bethesda, USA |
| *Valerie E. Jones* | Research, Enterprise & Development, Manchester Metropolitan University, UK |
| *Mai M. H. Mansour* | Department of Chemistry & Yousef Jameel Science & Technology Research Center, the American University In Cairo, Cairo, Egypt |
| *Garry McDowell* | Edge Hill University, Ormskirk L39 4QP, UK |
| *Dong M. Shin* | Winship Cancer Institute of Emory University Website Directions, Atlanta, GA 30322, United States |
| *Debatosh Majumdar* | Winship Cancer Institute of Emory University Website Directions, Atlanta, GA 30322, United States |
| *Xiang-Hong Peng* | Departments of Biomedical Engineering and Chemistry, Emory University and Georgia Institute of Technology, Atlanta |
| *May Azzawi* | School of Biology, Chemistry and Health Science, Manchester Metropolitan University, Manchester, UK |
| *Sylvain Martel* | École Polytechnique de Montréal (EPM) Campus of the University of Montréal, Montréal (Québec), Canada |
| *Kristen M. Jaruszewski* | Division of Basic Pharmaceutical Sciences, Florida A&M University, Tallahassee, FL 32307, USA |
| *Rajesh S. Omtri* | Division of Basic Pharmaceutical Sciences, Florida A&M University, Tallahassee, FL 32307, USA |
| *Karunya K. Kandimalla* | Division of Basic Pharmaceutical Sciences, Florida A&M University, Tallahassee, FL 32307, USA |

2

## CHAPTER 1

# Potential of Nanotechnology in Vascular Imaging and Treatment of Atherosclerosis

## Mark Slevin[*]

*Manchester Metropolitan University, Manchester, United Kingdom*

**Abstract:** Activation of vasa vasorum at specific sites in the adventitia initiates their proliferation or "angiogenesis" concomitant with development of atherosclerotic plaques. Haemorrhagic, leaky blood vessels from unstable plaques proliferate abnormally, are of relatively large calibre but are immature neovessels poorly invested with smooth muscle cells and possess structural weaknesses which may contribute to instability of the plaque by facilitation of inflammatory cell infiltration and haemorrhagic complications. Weak neovascular beds in plaque intima as well as activated adventitial blood vessels are potential targets for molecular imaging and targeted drug therapy, however, the majority of potential imaging and therapeutic agents have been unsuccessful because of their limited capacity to reach and remain stably within the target tissue or cells *in vivo*. Nanoparticle technology together with magnetic resonance imaging has allowed the possibility of imaging of neovessels in coronary or carotid plaques, and infusion of nanoparticle suspensions using infusion catheters or implant based drug delivery represents a novel and potentially much more efficient option for treatment. This review will investigate the possibility of future design of nanoparticles which home to the vasa vasorum and which can release siRNA directed against primary key external mitogens and intracellular signalling molecules in endothelial cells responsible for their activation with a view to inhibition of angiogenesis.

**Keywords:** Atherosclerosis; angiogenesis; plaque; cardiovascular disease; stroke; nanoparticle, targeting; imaging, microvessel; inflammation.

## INTRODUCTION

According to a World Health Organisation Fact Sheet (EURO/03/06) cardiovascular disease (CVD) is the number one killer in Europe, with heart disease and stroke being the major cause of death in all 53 Member States. Figures show that 34,421 (23% of all non-communicable diseases) of Europeans died from CVD in 2005. The report also highlighted the fact that there is approximately a 10-fold difference in premature CVD mortality between Western Europe and countries in Central and Eastern Europe (*i.e.* there is a higher occurrence of CVD amongst the poor and vulnerable). The problem for the European Union is that there is a direct correlation between the premature death rate and the viability of countries' economies. Although improvements in understanding have helped to reduce the number of Western European dying from CVD and related diseases further advances will require a clearer understanding of the pathobiological mechanisms responsible for the development of stroke, atherosclerosis and myocardial infarction. Approximately 75% of acute coronary events and 60% of symptomatic carotid artery disease are associated with disruption of atherosclerotic plaques [1]. As early as 1971, Folkman [2] introduced the concept of angiogenesis as a necessity for tumour growth. Its importance in other major pathological conditions, including, atherosclerosis, myocardial infarction and stroke was later realised [3, 4].

## 1. ATHEROSCLEROSIS AND NEOVESSEL FORMATION

Atherosclerosis is primarily a chronic inflammatory disorder; however, *angiogenesis* plays an important and complex role in development of unstable lesions [4]. During plaque development many pro-angiogenic pathways are re-acted and this leads to formation of immature blood vessels prone to rupture. Infiltration of microvessels into the media, intima and plaques, originates predominantly from proliferating vasa

*Address correspondence to Mark Slevin:* Manchester Metropolitan University, Manchester, United Kingdom; Tel: (0161) 2471172; Fax: (0161)2476365; E-mail: m.a.slevin@mmu.ac.uk

vasorum. Intima and media of coronary atherosclerotic vessels are infiltrated with a tumor-like mass of microvessels prone to leak [5]. However, chronic minimal injury also leads to intraluminal endothelial dysfunction and pro-inflammatory intracellular signalling pathways are recruited which lead to transcriptional up regulation of expression of cytokines, adhesion molecules and chemoattractant proteins.

Plaque angiogenesis is now accepted to have a fundamental role in the pathophysiological development of atherosclerosis, providing nutrients to the developing and expanding intima and also potentially creating an unstable haemorrhagic environment prone to rupture. The expression of intimal neovessels is directly related to the stage of plaque development, the presence of symptomatic disease and the risk of plaque rupture. In atherosclerosis, intimal neovascularization arises most frequently from the dense network of vessels in the adventitia, adjacent to a plaque, rather than from the main artery lumen. New blood vessels may have an active role in plaque metabolic activity and actively promote its growth beyond the critical limits of diffusion from the artery lumen. A strong correlation between areas of increased vascularity and intraplaque haemorrhage was first demonstrated by histological staining with anti-CD34 in symptomatic patients following endarterectomy [6]. The irregular nature of blood vessel formation has been likened to tumour angiogenesis, and hence the factors responsible for their growth may be different from those seen during normal wound healing. Our previous studies and those of others have suggested that haemorrhagic, leaky blood vessels from unstable carotid plaques proliferate abnormally. These relatively large calibre but immature neovessels are poorly invested with smooth muscle cells and possess structural weaknesses which may contribute to instability of the plaque by facilitation of inflammatory cell infiltration and haemorrhagic complications [7]. In a study of coronary artery atherogenesis, from patients subjected to heart transplant, lesions with the highest neovessels content were demonstrated to be of type VI and associated with the highest rate of thrombotic episodes [8].

These processes are in part initiated by hypoxia generated in the plaque, the specific action of growth factors, particularly Vascular Endothelial Growth Factor (VEGF) and basic Fibroblast Growth Factor (FGF-2), secreted by vascular and inflammatory cells and other factors such as hemodynamic stress. Immature neovessels may contribute to instability of the plaque by facilitation of inflammatory cell infiltration and haemorrhagic complications. Intraplaque haemorrhage results in rapid expansion of the plaque necrotic core, due to the fact that red blood cell membranes are a rich source of free cholesterol and phospholipids and the process occurring in association with excessive macrophage infiltration. The size of the necrotic core directly correlates with the risk of plaque rupture. Furthermore, intraplaque haemorrhage and plaque rupture were found to be proportional to neovessel density in coronary atheroma. Adventitial vessels in unstable plaques contain perivascular smooth muscle cells, however, after plaque rupture, the fibrous cap is disrupted with a luminal thrombus and the newer branches of vasa vasorum close to the necrotic core consist almost entirely of a single layer of EC overlying a ruptured, leaky basement membrane, and associated with remnants of red blood cells [1]. Defects are thought to be caused by proteolytic damage from on-going inflammation and release of signalling molecules affecting cell-cell contact.

There is promotion of wound healing through enhanced expression of a variety of growth factors in the neointima of the plaque. The sub endothelial world becomes extremely heterogeneous in composition. All the above support the hypothesis that angiogenesis plays a major role in the development of coronary and carotid artery symptomatic disease [9, 10]. Therefore, inhibition of angiogenesis might be an important target for prevention of development of active, unstable plaque lesions. Identification of pathophysiological changes related to angiogenesis may lead in future to the design of novel therapeutic agents. A number of candidate molecules and targets have already been identified. In the first place, important components of the plaque which are thrombogenic include fibrinogen, fibrin, FDP, and thrombin. Beyond the activity of thrombin in generating a fibrin clot and activating platelets, thrombin, affects endothelial cell migration and angiogenesis [11]. Other pro-angiogenic factors are found in these plaques *i.e.*: VEGF, Placental Growth Factor (PLGF), FGF-2, Transforming Growth Factor-Beta (TGF-β), Matrix Metalloproteinases (MMPs), Nitric Oxide (NO), Platelet-Derived Growth Factor (PDGF), Interleukin-8 (IL-8), and Platelet Activating Factor (PAF). As the thickness of the intima/media increases, the diffusion capacity of oxygen and nutrients from the lumen is exceeded. An angiogeneic response is stimulated by hypoxia and ischemia. Regulation of angiogenesis by hypoxia and role of the Hypoxia-Inducible Factor (HIF) system has been shown to be a major inducer of VEGF gene transcription [12].

An increasing number of angiogenic therapeutic targets have been proposed in order to facilitate modulation of neovascularization and its consequences in diseases such as cancer and macular degeneration. A complete knowledge of the mechanisms responsible for initiation of adventitial vessel proliferation, their extension into the intimal regions and possible de-novo synthesis of neovessels following differentiation of bone-marrow-derived stem cells is required in order to contemplate potential single or combinational anti-angiogenic therapies. Novel technologies employing laser-capture microdissection, RNA/DNA amplification and global genomic analysis can be used to identify optimal targets both extra and intra cellularly following isolation of individual primarily activated vessels [13].

## 2. NANOMEDICINE

Nanomedicine is the medical use of molecular-sized particles to deliver drugs, heat, light or other substances to specific cells in the human body. Engineering particles to be used in this way allows detection and/or treatment of diseases or injuries within the targeted cells, thereby minimizing the damage to healthy cells in the body. Nanotechnology exploits novel properties of materials when they are reduced to the size of a few hundred to thousand atoms. At this scale (anything from a few nanometres up to a hundred nanometres; a nanometre being one billionth of a metre) materials starts to exhibit quite different properties than would normally be expected. For example, gold appears red (the Romans used such nanoparticles of gold to stain glass); while titanium dioxide and zinc oxide (both used in sun-blocks) become transparent instead of white, while retaining their ability to block UV light. These are simple examples, however nanomaterials can have quite different physical (*e.g.*, strength, flexibility, thermal), electronic, magnetic, and optical properties compared with bulk materials. Some nanomaterials such as carbon nanotubes possess several different properties including strength (50-100 times stronger than steel), electronic properties (for example in displays), and biomedical uses (as drug delivery systems). As a result of these new properties, nanomaterials have the potential to impact every technology sector.

> *For medicine, nanotechnology promises new therapies, more rapid and sensitive diagnostic and investigative tools for normal and diseased tissues, and new materials for tissue engineering.*

Targeting therapeutic drugs directly at disease sites would increase effectiveness and reduce side effects. A range of nanoparticles (around 150nm in diameter) made of the biodegradable polymer poly(d,l-lactic-*co*-glycolic acid) and poly(ethylene glycol) have been developed which can encapsulate drugs, specific proteins or antibodies or which can be attached to artificial RNA strands known as aptamers. The nanoparticles can be taken up by specific cells where the nanoparticles dissolve to release the protein or drug. This technology has already been used to effectively treat and eliminate tumours in animal models [14]. The treatment of disease depends on the identification of a target and delivery of a therapeutic agent which either causes function to be restored, switches off inappropriate activity, or in the case of cancer, destroys the cell. However, many pharmaceuticals are limited in their development or application because of poor solubility, poor stability, or side-effects in inappropriate tissues. Manipulating the composition of a drug formulation at the nanoscale can resolve some of these issues. For example, various pharmaceutical companies in collaboration with researchers have shown that drugs can be stabilised at room temperature and ambient moisture when formulated as part of a nanostructured lattice with peptides and sugars. Similarly, in relation to use as possible treatments in disease, the anti-cancer drug paclitaxel was formulated with the human serum protein to create a nanostructure that requires no solvent, showed increased efficacy and consequently decreased side-effects (this is now licensed in the EU).

Although at an early stage of development, nanoshells can be used to encapsulate drugs, protecting them from the environment and offering targeted release (essential for toxic anti-cancer therapies). Nanoshells can be made from polymers (which fuse with cell membranes and release their contents within the cell) or a mix of polymer and gold (which can be induced to melt when irradiated with infrared light thus releasing their contents). Other nanomaterials can be used to target or directly treat diseased tissues by physical rather than biochemical means. For example, paramagnetic iron nanoparticles can be made to accumulate in tumour cells through the use of magnetic fields [15] and can be used to destroy tumour cells by heat through the application of alternating magnetic fields (known as Magnetic Fluid Hyperthermia [MFH]; [16, 17]. In the future functionalising such nanoparticles with targeting biomolecules could enable them to be delivered systemically.

Various types of nanoparticles have been considered for and used effectively in the field of medicine. For example, *Cantilever Arrays*- microscale levers which are functionalised at their ends with reference biomolecules. These bind appropriate target molecules in the sample causing the cantilever to bend (much like a diving board; [18]; *Carbon Nanotubes* (CNT) and fullerenes- essentially tubes and balls of carbon atoms, these can be used as porous cages to stabilise proteins. This enclosure effectively concentrates the protein and can also increase its specific activity, leading to an increased detection sensitivity of target molecules (down to pM levels; [19]; *Quantum Dots*- nanoscale semiconductor crystals that emit light of specific wavelengths dictated by their physical size: smaller dots fluoresce at shorter wavelengths, such as blue, while larger dots emit longer wavelengths, like red [20]; and *Membrane-based and Thin-film Sensors*- these allow the use of hydrophobic detector molecules (*e.g.* cell membrane proteins) or the compartmentalisation of detector molecules in a defined electrolytic environment (different from that of the sample; [21]. For molecular imaging, PFC nanoparticles can carry very large payloads of gadolinium to detect pathological biomarkers with Magnetic Resonance Imaging (MRI). Beads can be injected intravenously/intraventricularly followed by needle withdrawal, and the signal is created *via* the interaction between the water signal (proton density) and the magnetic properties, $R_1$ relaxation rate and $R_2$ transverse relaxation rate of the imaged tissues (a detailed description of this process is beyond the scope of this review but can be found in [22]). A variety of different epitopes, including alpha(v)beta(3)-integrin, tissue factor and fibrin, have been imaged using nanoparticles formulated with appropriate antibodies or peptidomimentics as targeting ligands. Lipophilic drugs can also be incorporated into the outer lipid shell of nanoparticles for targeted delivery. Upon binding to the target cell, the drug is exchanged from the particle surfactant monolayer to the cell membrane through a novel process called 'contact facilitated drug delivery'. By combining targeted molecular imaging and localized drug delivery, PFC nanoparticles provide diagnosis and therapy with a single agent [23].

## 3. NANOTECHNOLOGY APPLIED TO VISUALIZATION OF UNSTABLE PLAQUES AND ANGIOGENESIS IN CARDIOVASCULAR DISEASE

The majority of potential therapeutic agents have been unsuccessful because of their limited capacity to reach and remain stably within the target tissue or cell *in vivo*. Infusion of nanoparticle suspensions using infusion catheters or implant based drug delivery represents a novel and potentially much more efficient option for treatment [24]. As mentioned previously, neovascular beds in plaque intima as well as activated adventitial blood vessels are potential targets for molecular imaging and targeted drug therapy. To achieve effective molecular targeting and imaging, the particles must be designed to have a long circulating half-life, to be sensitive and selective to the epitope of interest and produce a prominent contrast to noise ratio enhancement. Obviously they should also be non-toxic. Fibrin-targeted nanoparticles can enhance the contrast in thrombi, as demonstrated in the femoral and carotid arteries of a miniswine model of atherosclerosis (injury-high-cholesterol) using echogenic immunoliposome targeted anti-fibrin antibodies [25, 26]. *Perfluorocarbon* emulsions of 200-300nm have been used in conjunction with Positron Emission Tomography (PET) and radio-labelled antibodies for imaging of atherosclerotic plaque angiogenesis [27], and fibrin-specific perfluorocarbon particles were able to deliver effectively plasminogen activator streptokinase in human plasma clots suggesting a role for nano-based targeted thrombolysis in patients with symptomatic cardio/cerebro vascular disease [28].

When large numbers of *paramagnetic gadolinium* complexes (>50,000) are incorporated onto emulsion particles, the signal enhancement for each binding site is magnified dramatically compared with conventional contrast reagents [29] and references therein). Neovessels in coronary plaques of cholesterol-fed apoE mice have been imaged using Vascular Cell Adhesion Molecule-1 Peptide Sequence (VHSPNKK) bound to cross-linked magneto-fluorescent **super-paramagnetic iron oxide** (CLIO) particles. The particles were shown to selectively bind to aortic plaque vasculature 24h after injection using MRI and *ex vivo* MR and corresponded with histology and fluorescent analysis of tissue samples performed after euthanasia, suggesting potential diagnostic and therapeutic applications [30]. Tissue factor is a prothrombotic molecule which becomes exposed during plaque rupture and represents a key molecule associated with plaque destabilization Ultra-small iron oxide particles, prepared as a liquid perfluorocarbon contrast agent with high gadolinium payload (92,400/bead) were conjugated to anti-tissue factor antibodies and following injection at pico-molar concentrations, MRI at 1.5T was able to quantify expression in smooth muscle cell monolayer cultures. [16]. *In vivo* injection of quantum dots and/or fluorophore-labelled markers can help to elucidate their expression in living models following *ex vivo* measurements, whilst targeted delivery of antisense molecules using the same technology could modulate their expression and ability to activate local microvasculature, although this has yet to be attempted to our knowledge for the treatment of atherosclerosis [31].

## SUMMARY

These contrast reagents are an exiting development in imaging of atherosclerosis and represent a potential technology in association with MRI capable of visualising plaque features associated with instability as well as identifying total plaque burden. As mentioned earlier, a number of markers of angiogenesis have been identified which might lend themselves to this new technology and help in identifying unstable haemorrhagic regions of plaques. The $\alpha_5\beta_3$ integrin is a membrane-bound adhesion molecule widely expressed, particularly by active EC (not quiescent ones) and smooth muscle cells associated with new blood vessel formation. [32] demonstrated that these integrin-targeted nanoparticles could detect and characterise angiogenesis patterns associated with atherosclerosis in the vasa vasorum of coronary arteries from cholesterol-fed rabbits.

Future treatments and/or preventative measures may involve synthesis of superparamagnetic nanoparticles multiply/quadruple labeled with antibodies to CD105 or integrins which bind selectively with active microvessels, siRNA directed to key early mitogenic activators and/or intracellular signaling molecules (yet to be defined as part of a blue print for EC activation), Green Fluorescent Protein (GFP) to enable near infrared *in vivo* optical imaging (NIRF) and confirm probe delivery, and Myristoylated Polyarginine Peptides (MPAP) to allow membrane translocation of the siRNA. T2 relaxation times can be monitored to examine probe delivery. Whilst this type of combination therapy might be some years away, current studies should aim to form the platform by identifying specific activatory signals of vasa vasorum and intracellular signal transduction consequences (*i.e.* a blue print for activation) using currently available *in vivo* models combined with nanotechnology. Specifically directed siRNA expressed in adenoviral vectors and targeted with nanoparticles might be one method to prime the cells against self or external activation and by prevention of angiogenesis, significantly reduce the rate of formation of arterial plaque.

Nanoparticles could then be used for effective delivery of drugs to target sites with a preferential aim to prevent initial development of neointimal vascular sites by blocking adventitial blood vessel activation in patients with low grade plaques. With this in mind, novel specific markers of plaque angiogenesis such as CD105 [33], modified C-reactive protein [34] and VEGF-R2 [35] might represent 1) future imaging targets able to identify patients developing unstable plaque regions susceptible to rupture, and 2) prevent or slow down development of atheroma thereby improving treatment and survival rates of patients with a history of development of myocardial infarction or ischaemic stroke.

## REFERENCES

[1]    Kolodgie FD, Narula J, Yuan C. *et al.* Elimination of neoangiogenesis for plaque stabilization: is there a role for local drug therapy? J Am Coll Cardiol 2007;49:2093-3001.

[2]    Folkman J. Tumour angiogenesis: therapeutic implications. N Eng J Med 1971;285:1182-86.

[3]    Slevin M, Kumar P, Gaffney J, *et al.* Can angiogenesis be exploited to improve stroke outcome? Mechanisms and therapeutic potential. Clin Sci 2006;111:171-83.

[4] Slevin M, Wang Q, Angels Font M *et al.* Atherothrombosis and plaque heterology: different location or a unique disease? Pathobiol 2008;75:209-225.

[5] Barger AC, Beeuwkes R, 3rd, Lainey LL, *et al.* Hypothesis: vasa vasorum and neovascularization of human coronary arteries. A possible role in the pathophysiology of atherosclerosis. N Engl J Med 1984;310:175-77.

[6] Mofidi R, Crotty TB, McCarthy P, *et al.* Association between plaque instability, angiogenesis and symptomatic carotid occlusive disease. Br J Surj 2001;88: 945-50.

[7] Dunmore BJ, McCarthy MJ, Naylor AR, *et al.* Carotid plaque instability and ischaemic symptoms are linked to immaturity of microvessels within plaques. J Vasc Surg 2007;45: 155-59.

[8] Juan-Babot JO, Martínez-González J, Berrozpe M, *et al.* Neovascularization in human coronary arteries with lesions of different severity. Rev Esp Cardiol 2003;56:978-86.

[9] O'Brien KD, Chait A. The biology of the artery wall in atherogenesis. Med Clin North Am 1994;78:41-67.

[10]. McCarthy MJ, Loftus IM, Thompson MM, *et al.* Angiogenesis and the atherosclerotic carotid plaque: an association between symptomatology and plaque morphology. J Vasc Surg 1999;30:261-68.

[11] Zania P, Papaconstantinou M, Flordellis CS, *et al.* Thrombin mediates mitogenesis and survival of human endothelial cells through distinct mechanisms. Am J Physiol 2008;294:1215-26.

[12] Pugh CW, Ratcliffe PJ. Regulation of angiogenesis by hypoxia: role of the HIF system. Nat Med 2003;9:677-684.

[13] Roy A, Patel D, Khanna S, *et al.* Transcriptome-wide analysis of blood vessels laser captured from human skin and chronic wound-edge tissue. PNAS 2007;104:14472-14477.

[14] Kim JH, Kim YS, Park K, *et al.* Antitumor efficacy of cisplatin-loaded glycol chitosan nanoparticles in tumor-bearing mice. J Control Release 2007;127: 41-49.

[15] Alexiou C, Schmid RJ, Jurgons R, *et al.* Targeting cancer cells: magnetic nanoparticles as drug carriers. Eur Biophys J 2006;35: 446-450

[16] Morawski AM, Winter PM, Crowder PM, *et al.* Targeted nanoparticles for quantitative imaging of sparse molecular epitopes with MRI. Magn Reson Med 2004;51:480-486.

[17]. Johannsen M, Gneveckow U, Thiesen B, *et al.* Thermotherapy of prostate cancer using magnetic nanoparticles: feasibility, imaging, and three-dimensional temperature distribution. Eur Urol 2007;52:1653-1661.

[18] Calleja M, Nordström M, Alvarez M, *et al.* Highly sensitive polymer-based cantilever-sensors for DNA detection. Ultramicroscopy 2005;105:215-222.

[19] Sotiropoulou S, Gavalas V, Vamvakaki V, *et al.* Novel carbon materials in biosensor systems. Biosens Bioelectron 2003;18:211-215.

[20] Riegler J, Nann T. Application of luminescent nanocrystals as labels for biological molecules. Anal Bioanal Chem 2004;379:913-919.

[21] Yakimova R, Steinhoff G, Petoral RM Jr, *et al.* Novel material concepts of transducers for chemical and biosensors. Biosens Bioelectron 2007;22:2780-2785.

[22] Waters EA, Wickline SA. Contrast agents for MRI. Basic Res Cardiol 2008;103:114-121.

[23] Winter PM, Cai K, Caruthers SD, *et al.* Emerging nanomedicine opportunities with perfluorocarbon nanoparticles. Exp Rev Med Devices 2007;4:137-145.

[24] Patel DN. Nanotechnology in cardiovascular medicine. Cath Cardiovasc Interven 2007;69:643-654.

[25]. Hamilton AJ, Huang SL, Warnick D, *et al.* Intravascular ultrasound molecular imaging of atheroma components *in vivo*. J Am Coll Cardiol 2004;43:453-460.

[26] Morawski AM, Lanza GA, Wickline SA. Targeted contrast agents for magnetic resonance imaging and ultrasound. Curr Opin Biotechnol 2005;16:89-92.

[27] Warburton L, Gillard J. Functional imaging of carotid atheromatous plaques. J. Neuroimaging 2006;16:293-301.

[28] Marsh JN, Senpan A, Hu G, *et al.* Fibrin-targeted perfluorocarbon nanoparticles for targeted thrombolysis. Nanomed 2007;2:533-543.

[29] Wickline SA, Neubauer AM, Winter PM, *et al.* Molecular imaging and therapy of atherosclerosis with targeted nanoparticles. J Mag Res Imag 2007;25: 667-680.

[30] Kelly KA, Allport JR, Tsourkas A, *et al.* Detection of vascular adhesion molecule-1 expression using a novel multimodal nanoparticle. Circ Res 2005;96:327-336.

[31] Medarova Z, Pham W, Farrar C, *et al. In vivo* imaging of siRNA delivery and silencing in tumours. Nature Med 2007;13:372-377.

[32] Winter PM, Caruthers SD, Yu X, *et al.* Improved molecular imaging contrast agent for detection of human thrombus. Mag Res Med 2003;50: 411-416.

[33]  Li C, Mollahan P, Baguneid MS, *et al.* A comparative study of neovascularisation in atherosclerotic plaques using CD31, CD105 and TGF beta 1. Pathobiol 2006;73:192-197.

[34]  Krupinski J, Turu MM, Martinez-Gonzalez J, *et al.* Endogenous expression of C-reactive protein is increased in active (ulcerated noncomplicated) human carotid artery plaques. Stroke 2006;37: 1200-1204.

[35]  Hauer AD, van Puijvelde GH, Peterse N, *et al.* Vaccination against VEGFR2 attenuates initiation and progression of atherosclerosis. Arterioscler Thromb Vasc Biol 2007;27:2050-2057.

## CHAPTER 2

# Effective Transvascular Delivery of Chemotherapy into Cancer Cells with Imageable Nanoparticles in the 7 to 10 Nanometer Size Range

## Hemant Sarin[*]

*National Institute of Biomedical Imaging and Bioengineering, National Institutes of Health, Bethesda, MD, USA*

**Abstract:** The physiologic upper limit of pore size in the blood-tumor barrier of cancer microvasculature is approximately 12 nanometers, independent of whether cancer location is in the brain and the central nervous system, or outside, in peripheral tissues. Chemotherapy drugs in clinical use are less than 1 to 2 nanometers in diameter and can readily extravasate across the blood-tumor barrier to enter the extravascular compartment of cancer tissue. However, these small molecule chemotherapy drugs maintain peak blood concentrations for only a few minutes, and therefore, do not accumulate to high concentrations within individual cancer cells in the extravascular compartment. Spherical nanoparticles in the 7 to 10 nanometer size range maintain peak blood concentrations for several hours and are sufficiently smaller than the 12 nm physiologic upper limit of pore size within the blood-tumor barrier to accumulate to high concentrations within individual cancer cells. The Gd-G5-doxorubicin dendrimer is an imageable nanoparticle bearing chemotherapy within the 7 to 10 nanometer size range. Doxorubicin attachment to the Gd-G5-doxorubicn dendrimer *via* the pH-sensitive covalent hydrazone bond facilitates efficient intracellular release of doxorubicin and doxorubicin accumulation in cancer cell nuclei. One dose of the Gd-G5-doxorubicin dendrimer is significantly more effective than one dose of free doxorubicin at inhibiting the growth of RG-2 rodent brain cancers for 24 hours. The therapeutic efficacy of the Gd-G5-doxorubicin dendrimer *in vivo* stems from the effective transvascular delivery of doxorubicin across the blood-tumor barrier into individual brain cancer cells and doxorubicin accumulation to high concentrations within brain cancer cell nuclei. It is foreseeable that such imageable nanoparticles bearing chemotherapy, which are within the 7 to 10 nanometer size range, will also demonstrate therapeutic efficacy in the treatment of cancers located outside the brain and central nervous system.

**Keywords:** Blood brain barrier; cancer; chemotherapy; nanoparticles; doxorubicin; dendrimer; magnetic resonance imaging; tumour cells, microvascular; gadolinium.

## 1. BACKGROUND

Simple diffusion of nutrients and metabolites between cancer cells and the pre-existent microvasculature of host tissues is only sufficient to sustain cancer growth to a volume of 1 to 2 mm$^3$ [1]. Further additional cancer growth is contingent on the formation of new microvasculature within cancer tissue, a process that is mediated by Vascular Endothelial Growth Factor (VEGF) [2, 3]. The new microvasculature induced by VEGF is of the discontinuous type due to the existence of anatomic defects within and in between endothelial cells of the Blood-Tumor Barrier (BTB) [3, 4]. For this reason, the pathologic BTB of cancer microvasculature is leakier than the endothelial cell barriers of microvasculature supplying most normal healthy tissues [5-13]. The anatomic defects in the BTB may be several hundred nanometers wide at the ultrastructural level [14-16], however, in the physiologic state *in vivo*, the fibrous luminal glycocalyx layer would be present over the anatomic defects in the BTB [12, 13, 17]. Therefore, the physiologic upper limit of pore size in the BTB is well-defined, and is approximately 12 nm, independent of cancer location being in the brain and the central nervous system, or outside, in peripheral tissues [12, 13, 17].

The 12 nm physiologic upper limit of pore size in the BTB of cancer microvasculature is not a significant impediment to the transvascular passage of systemically administered chemotherapy drugs across the BTB

*Addres correspondence to Hemant Sarin: National Institute of Biomedical Imaging and Bioengineering, National Institutes of Health, Bethesda, MD, USA; Tel: 202 865 3726; E-mail: hemantsarin74@gmail.com

**Mark Slevin (Ed)**

into the extravascular compartment of cancer tissue [11, 18-20], since most chemotherapy drugs are small molecules with molecular weights less than 1 kDa and diameters less than 1 to 2 nm. This includes classical chemotherapy drugs that target the cell cycle, such as DNA alkylating drugs [21, 22], and newer investigational drugs that target cell surface receptors and associated pathways, such as the tyrosine kinase inhibitors [23]. Chemotherapy drugs have been ineffective at treating cancer [24-27], since these small molecule drugs are rapidly eliminated from systemic blood circulation by the kidneys, which efficiently filter particles smaller than 5 to 6 nm in diameter [10, 28]. Chemotherapy drugs only maintain peak blood concentrations for a few minutes after each dose, and therefore, do not accumulate to therapeutic concentrations within individual cancer cells [11, 13].

Over the past few decades, various slow sustained-drug release formulations of small molecule chemotherapy drugs have been developed by non-covalently attaching drugs to polymers (*i.e.* albumin-bound paclitaxel), or alternatively, by encapsulating drugs in liposomes (*i.e.* liposomal doxorubicin), for slow continuous drug release over time [29, 30]. Since the particle sizes of slow sustained-drug release formulations typically range between 50 nm and 250 nm in diameter [31, 32], these nanoparticles are significantly larger than the 12 nm physiologic upper limit of pore size within the BTB of tumor microvasculature [12, 13], and as a consequence, particles are too large to extravasate across the BTB of cancer microvasculature. Therefore, nanoparticles larger than the 12 nm physiologic upper limit of pore size within the BTB are 'intravascular drug reservoirs" from which free drug is slowly released into systemic blood circulation. Since the free drug released into circulation from these large nanoparticles itself has a short blood half-life, it does not accumulate to high concentrations within individual tumor cells [13, 17].

It has been recently shown that imageable dendrimer nanoparticles in the 7 to 10 nm size range maintain peak blood concentrations for several hours and are sufficiently smaller than the 12 nm physiologic upper limit of pore size within the BTB of cancer microvasculature to accumulate to high concentrations within individual cancer cells [12, 13, 17]. The covalent attachment of chemotherapy drugs to such nanoparticles *via* labile bonds allows for efficient release of drug following nanoparticle endocytosis, so that drug fraction can accumulate to effective concentrations in cancer cell nuclei [17]. The Gd-G5-doxorubicin dendrimer is an imageable nanoparticle bearing chemotherapy within the 7 to 10 nm size range with doxorubicin attached *via* a pH-sensitive hydrazone bond [17], an acid labile covalent bond from which doxorubicin is efficiently released following particle endocytosis [33-35]. In RG-2 rodent brain cancer regression studies, it has been shown that one dose of the Gd-G5-doxorubicin dendrimer is significantly more effective than one dose of free doxorubicin at inhibiting the growth of RG-2 rodent brain cancers for 24 hours [17]. The therapeutic efficacy of the Gd-G5-doxorubicin dendrimer stems from the effective transvascular delivery of doxorubicin across the BTB of brain cancer microvasculature into individual brain cancer cells and doxorubicin accumulation to high concentrations within brain cancer cell nuclei [17]. In this chapter, I will discuss this recent translational research on the *in vivo* use of imageable nanoparticles bearing chemotherapy that are within the 7 to 10 nm size range for effective transvascular drug delivery across the BTB into cancer cells.

## 2. DENDRIMER NANOPARTICLES

Dendrimers are particularly small multigenerational nanoparticles that are spherical, and possess surface groups that can be functionalized with small molecule imaging, targeting, and therapeutic agents [36-38]. Dendrimer size increases by only 1 to 2 nm in diameter with each successive generation, while the number of dendrimer surface groups doubles with each successive generation. For example, polyamidoamine (PAMAM) dendrimers functionalized with gadolinium (Gd)-diethyltriaminepentaacetic acid (DTPA), a small molecule MRI contrast agent, range in diameter between 1.5 nm (Gd-DTPA PAMAM dendrimer generation 1, Gd-G1) and 14 nm (Gd-DTPA PAMAM dendrimer generation 8, Gd-G8) [12, 13, 38, 39]. The molecular weight increase of the naked dendrimer to that of the Gd-DTPA conjugated dendrimer is proportional to the percent conjugation of Gd-DTPA (Table **1**) [13].

**Table 1.** Physical properties of naked and Gd-DTPA conjugated PAMAM dendrimer generations. Adapted from reference [13].

| PAMAM dendrimer generation (G) | Terminal amines (#) | Naked PAMAM dendrimer molecular weight (kDa) | Gd-DTPA conjugated PAMAM dendrimer molecular weight (kDa) | Gd-DTPA conjugation (%) |
|---|---|---|---|---|
| G1 | 8 | 1.4 | 5.6 | 67 |
| G2 | 16 | 3.3 | 11 | 66 |
| G3 | 32 | 6.9 | 19 | 48 |
| Lowly conjugated G4 | 64 | 14 | 24 | 30 |
| Standard G4 | 64 | 14 | 40 | 48 |
| G5 | 128 | 29 | 80 | 47 |
| G6 | 256 | 58 | 133 | 40 |
| G7 | 512 | 116 | 330 | 50 |
| G8 | 1024 | 233 | 597 | 38 |

The number of surface groups doubles with each successively higher PAMAM dendrimer generation (Table 1; Fig. 1) [13]. The exterior of naked PAMAM dendrimers is positively charged due to the presence of terminal amine groups (Fig. 1) [13]. Since each Gd-DTPA carries a charge of -2, the conjugation of Gd-DTPA to a proportion of the terminal amine groups on PAMAM dendrimer exterior neutralizes the positively charged exterior of naked PAMAM dendrimers (Fig. 2) [13].

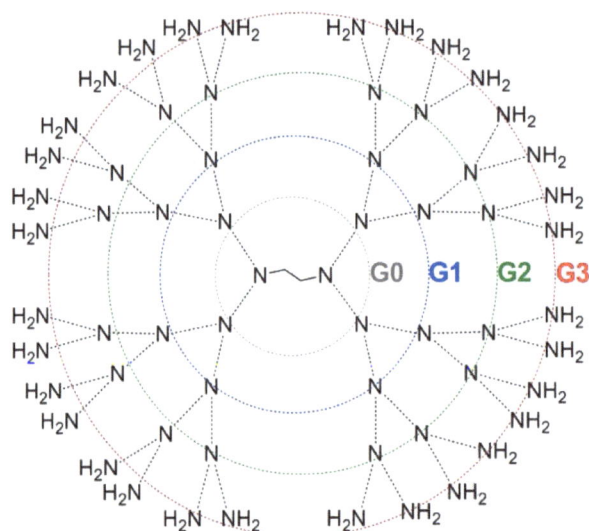

**Figure 1:** Illustration of naked PAMAM dendrimer generations from the ethylenediamine (EDA) core (G0) to generation 3 (G3). Adapted from reference [13].

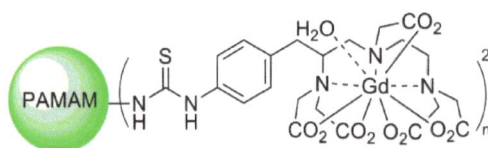

**Figure 2:** Gd-DTPA conjugated PAMAM dendrimer (Gd-dendrimer). Adapted from reference [13].

The masses of Gd-G5 through Gd-G8 dendrimer particles are sufficient enough for particle visualization by annular dark-field scanning transmission electron microscopy (ADF STEM) (Fig. **3**) [13]. The sizes of Gd-G7 and Gd-G8 dendrimer particles visualized by ADF STEM are large enough for measurement of particle diameters, which are approximately 11 nm for Gd-G7 dendrimers and approximately 13 nm for Gd-G8 dendrimers [12, 13].

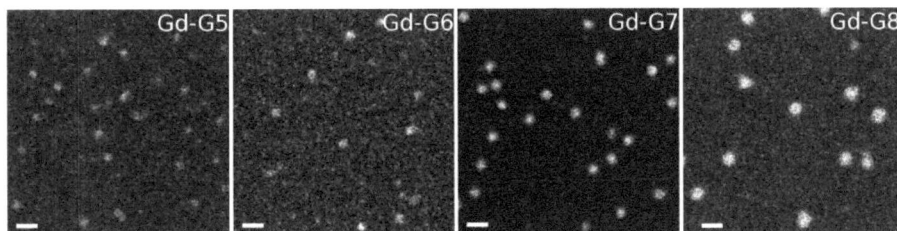

**Figure 3:** Annular dark-field scanning transmission electron microscopy (ADF STEM) images of Gd-G5, Gd-G6, Gd-G7, and Gd-G8 dendrimers adsorbed onto an ultrathin carbon support film. Scale bar = 20 nm. Adapted from reference [13].

## 3. SIGNIFICANCE OF POSITIVE CONTRAST ENHANCEMENT OF CANCER TISSUE ON MAGNETIC RESONANCE IMAGING

The ready transvascular passage of small molecule paramagnetic MRI contrast agents, such as Gd-DTPA (Magnevist; 0.938 kDa), across the porous BTB of cancer microvasculature results in temporary positive contrast enhancement of cancer tissue on $T_1$-weighted MRI [40-42], since the relaxation rate of water protons in the extravascular compartment increases when the contrast agent is present in the compartment [43]. With $T_1$-weighted dynamic contrast-enhanced MRI it is possible to repetitively re-image cancer tissue over time following an intravenous bolus of a contrast agent. Therefore, with dynamic contrast-enhanced MRI it is possible to measure the degree to which Gd-DTPA conjugated PAMAM dendrimers extravasate across the BTB of cancer microvasculature and accumulate in the extravascular compartment of cancer tissue [12, 13, 17]. The Gd-dendrimer concentration in cancer tissue can be estimated by the *in vivo* measurement of tumor tissue MRI signal at baseline ($T_{10}$) and then again following the intravenous infusion of the Gd-dendrimer ($T_1$), and the *in vitro* measurement of the molar relaxivity ($r_1$) of the Gd-dendrimer, which is the proportionality constant for conversion of Gd signal to Gd concentration (Eq. **1**) [12, 13, 17, 44].

$$\frac{1}{T_1} - \frac{1}{T_{10}} = r_1 \cdot [Gd] \tag{1}$$

## 4. THE PHYSIOLOGIC UPPER LIMIT OF PORE SIZE IN THE BLOOD-TUMOR BARRIER

The physiologic upper limit of pore size in the BTB of cancer microvasculature is approximately 12 nm, independent of whether cancer location is within the brain and central nervous system, or outside, in peripheral tissues [12, 13]. Furthermore, it appears that the physiologic upper limit of pore size in the BTB of small new cancer lesions with volumes less than 25 to 30 $mm^3$ is only 1 to 2 nm lower than the upper limit of pore size in the BTB of large well-developed cancer lesions [13].

The physiologic upper limit of pore size in the BTB of cancer microvasculature has been probed by dynamic contrast-enhanced MRI. In these dynamic contrast-enhanced MRI experiments successively higher generation Gd-DTPA conjugated PAMAM dendrimers were intravenously administered to rodents bearing RG-2 malignant gliomas. In one series of dynamic contrast-enhanced MRI experiments, Gd-G1 through Gd-G8 dendrimers were administered to rodents with RG-2 malignant gliomas growing orthotopically in brain tissue [13]. For these experiments the orthotopic RG-2 glioma model served as a representative model of brain cancer [45, 46]. In another series of experiments, Gd-G5 through Gd-G8 dendrimers were administered to rodents with RG-2 malignant gliomas growing orthotopically in brain tissue as well as ectopically in temporalis skeletal muscle tissue [12]. In these experiments, the RG-2 gliomas growing ectopically in skeletal muscle tissue served as a representative model of peripheral cancer [16, 47-49].

The degrees of Gd-G1 through Gd-G8 dendrimer extravasation across the BTB of RG-2 brain cancer microvasculature over time are evident on the Gd concentration maps from representative time points of 2-hour long dynamic contrast-enhanced MRI sessions following the intravenous boluses of the respective Gd-dendrimer generations over 1 minute. In the case of large RG-2 brain cancers (Fig. 4) [13], Gd-G1 through Gd-G7 dendrimers extravasate across the BTB into the extravascular compartment of brain cancer tissue, which is evident as positive contrast enhancement in brain cancer tissue. Gd-G8 dendrimers do not extravasate across the BTB of even large RG-2 cancers, but instead remain intravascular, which is evident as lack of contrast enhancement in brain cancer tissue (Fig. 4) [13]. Since the diameter of Gd-G8 dendrimers is approximately 13 nm and the diameter of Gd-G7 dendrimers is approximately 11 nm, the physiologic upper limit of pore size is approximately 12 nm in the BTB of large RG-2 brain cancers [12, 13].

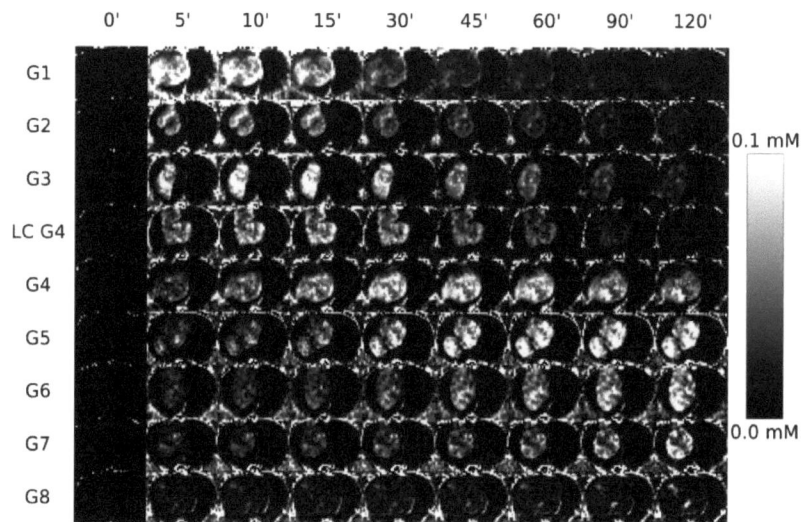

**Figure 4:** $T_1$-weighted dynamic contrast-enhanced MRI-based Gd concentration maps of Gd-dendrimer distribution over time within large RG-2 brain cancers.

The volume, in mm$^3$, for each large orthotopic RG-2 glioma: Gd-G1, 104; Gd-G2, 94; Gd-G3, 94; lowly conjugated (LC) Gd-G4, 162; standard Gd-G4, 200; Gd-G5, 230; Gd-G6, 201; Gd-G7, 170; and Gd-G8, 289. Respective Gd-dendrimer generations administered intravenously over 1 minute at a Gd dose of 0.09 mmol Gd/kg animal body weight. Scale from 0 mM [Gd] to 0.1 mM [Gd]. Adapted from reference [13].

In the case of small RG-2 brain cancers (Fig. 5) [13], only Gd-G1 through Gd-G6 dendrimers extravasate across the BTB into the extravascular compartment of small brain cancers. Gd-G7 dendrimers do not extravasate across the BTB of the microvasculature of small RG-2 brain cancers with volumes less than 25 to 30 mm$^3$ (Fig. 5) [13]. Therefore, the physiologic upper limit of pore size in the BTB of small RG-2 brain cancers is 1 to 2 nm lower than that in the BTB of large RG-2 brain cancers. This would be attributable to the presence of a thicker and more intact luminal glycocalyx layer of the BTB of small new cancer lesions [12, 17].

The volume, in mm$^3$, for each small orthotopic RG-2 glioma: Gd-G1, 27; Gd-G2, 28; Gd-G3, 19; lowly conjugated (LC) Gd-G4, 24; standard Gd-G4, 17; Gd-G5, 18; Gd-G6, 22; Gd-G7, 24; and Gd-G8, 107. Respective Gd-dendrimer generations administered intravenously over 1 minute at a Gd dose of 0.09 mmol Gd/kg animal body weight. Scale from 0 mM [Gd] to 0.1 mM [Gd]. Adapted from reference [13].

In cases of both large and small RG-2 brain cancers, Gd-G1 through standard Gd-G4 dendrimers only accumulate within the extravascular compartment of cancer tissue for a short period of time (Figs. 4 and 5) [13], as these lower Gd-dendrimer generations possess relatively short blood half-lives and maintain peak blood concentrations for only a few minutes, as particle sizes are smaller than 5 to 6 nm in diameter. The greater temporary accumulation of Gd-G1 through standard Gd-G4 dendrimers in the extravascular compartment of large RG-2 brain cancers (Fig. 4) [13], as compared to in the extravascular compartment of small RG-2 brain cancers (Fig. **5**) [13], is attributable to large RG-2 brain cancers being more vascular than small RG-2 brain cancers [11, 13].

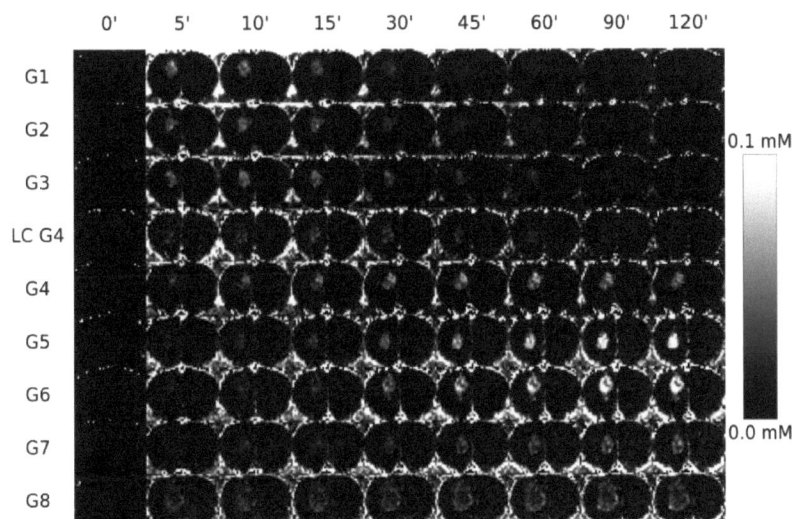

**Figure 5:** $T_1$-weighted dynamic contrast-enhanced MRI-based Gd concentration maps of Gd-dendrimer distribution over time within small rodent brain cancers.

Both Gd-G5 and Gd-G6 dendrimers readily extravasate across the BTB of small and large RG-2 brain cancers (Figs. **4** and **5**) [13]. Even through small RG-2 brain cancers are less vascular than large RG-2 brain cancers, Gd-G5 and Gd-G6 dendrimers effectively accumulate over time within the extravascular compartments of small and large RG-2 brain cancers (Figs. **4** and **5**) [13], since these generations possess relatively long blood half-lives and maintain peak blood concentrations for several hours, as particle sizes are within the 7 to 10 nm diameter size range.

The physiologic upper limit of pore size in the BTB of RG-2 peripheral cancer microvasculature is also approximately 12 nm [12], since Gd-G8 dendrimers are too large to cross the pores within the BTB of RG-2 peripheral cancer microvasculature (Fig. **6**) [12]. With each successively higher Gd-dendrimer generation from Gd-G5 to Gd-G8, there is progressively less particle extravasation across the BTB of both RG-2 peripheral cancer and brain cancer microvasculature (Fig. **6**) [12].

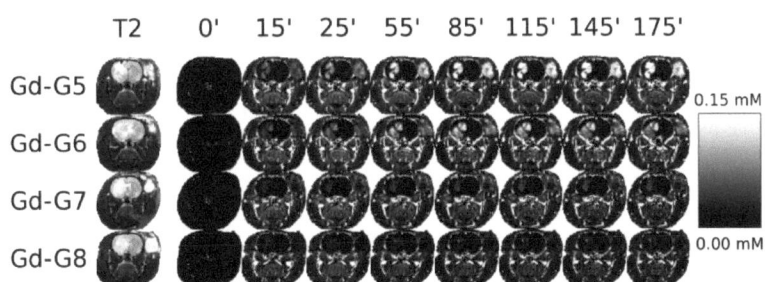

**Figure 6:** $T_2$-weighted anatomical MRI images and $T_1$-weighted dynamic contrast-enhanced MRI-based Gd concentration maps of high generation Gd-dendrimer distribution over time within RG-2 brain cancers as compared to RG-2 peripheral cancers.

The volume, in mm³, for each orthotopic and ectopic RG-2 glioma: First row, Gd-G5; 45 mm³ (brain), 113 mm³ (peripheral); Second row, Gd-G6; 97 mm³ (brain), 184 mm³ (peripheral); Third row, Gd-G7; 53 mm³ (brain), 135 mm³ (peripheral); Fourth row, Gd-G8; 50 mm³ (brain); 163 mm³ (peripheral). Respective Gd-dendrimer generations administered intravenously over 1 minute at a Gd dose of 0.09 mmol Gd/kg animal body weight. Scale from 0.00 mM [Gd] to 0.15 mM [Gd]. Adapted from reference [12].

The Gd-G8 dendrimers remain intravascular in both RG-2 peripheral cancers and brain cancers, and therefore, demonstrate similar pharmacokinetics over 600 to 700 minutes in both RG-2 peripheral cancers and brain cancers (Fig. **7**) [12]. Although the physiologic upper limit of pore size in the BTB of both RG-2 peripheral cancer and brain cancer microvasculature is equivalent, there is less transvascular extravasation of Gd-G5 through Gd-G7 dendrimers across the BTB of RG-2 brain cancers than across the BTB of RG-2 peripheral cancers (Fig. **7**) [12]. Since there are no differences in the physiologic upper limit of pore size within the BTB of both RG-2 peripheral and brain cancer microvasculature, this lesser transvascular extravasation of Gd-G5 through Gd-G7 dendrimers across the BTB of brain cancer microvasculature is likely attributable to the presence of fewer anatomic defects underlying the luminal glycocalyx of the BTB of brain cancer microvasculature [12].

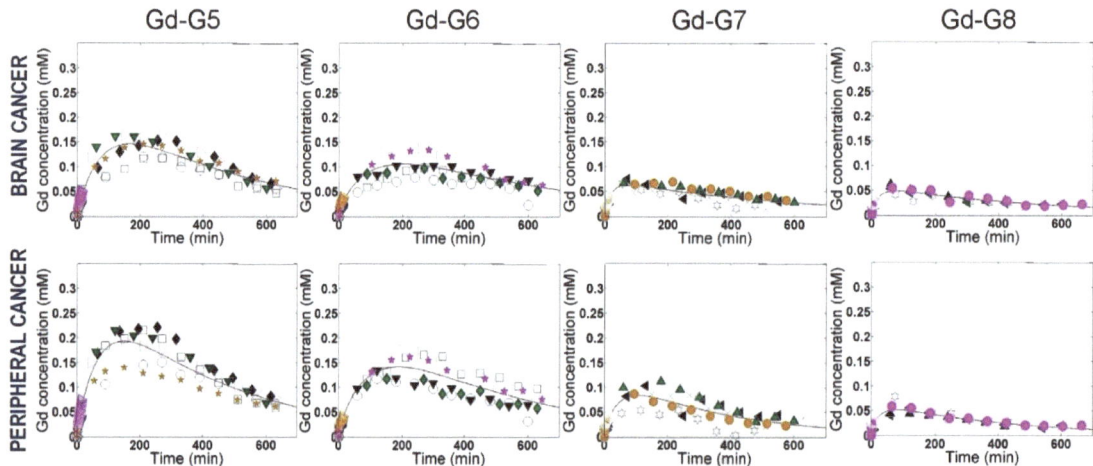

**Figure 7:** Dynamic contrast-enhanced MRI-based pharmacokinetics of high generation Gd-dendrimers over 600 to 700 minutes in RG-2 brain cancers as compared to RG-2 peripheral cancers.

Gd-G5 (n=6), Gd-G6 (n=6), Gd-G7 (n=5), Gd-G8 (n=5). Whole tumor tissue Gd concentrations for the RG-2 brain cancers and RG-2 peripheral cancers calculated for each of the dynamic scan session time-points. Respective Gd-dendrimer generations intravenously administered over 1 minute at a Gd dose of 0.09 mmol Gd/kg during the initial 15 minute dynamic contrast-enhanced MRI scan session. Subsequent dynamic scan sessions of re-anesthetized animals conducted at 30 to 90 minute time intervals. Adapted from reference [12].

## 5. SIGNIFICANCE OF THE LUMINAL GLYCOCALYX LAYER OF THE BLOOD-TUMOR BARRIER IN THE PHYSIOLOGIC STATE *IN VIVO*

The relatively thick fibrous polysaccharide meshwork of the luminal glycocalyx layer is the primary point of resistance to the transvascular passage of macromolecules across the endothelial cell barriers of microvasculature supplying normal and inflamed tissues in the physiologic state *in vivo* [50-52]. Therefore, the well-defined 12 nm physiologic upper limit of pore size in the BTB of cancer microvasculature would be attributable to the presence of a fibrous luminal glycocalyx layer over the anatomic defects within the BTB of cancer microvasculature in the physiologic state *in vivo* [12, 13, 17]. Furthermore, the presence of a slightly thicker and more intact glycocalyx layer over the defects in the BTB of small new cancer lesions is the likely reason for the slightly lower physiologic upper limit of pore size in the BTB of small new cancer lesions, as is the case in the BTB of small RG-2 brain cancers with volumes less than 25 to 30 mm$^3$ [13, 17].

Due to the presence of the luminal glycocalyx layer in the physiologic state *in vivo*, the underlying endothelial cells of the BTB are not accessible to liposomes, viruses, bacteria, and cells, unless the glycocalyx is stretched, degraded, or disrupted in some manner [53-56]. The glycocalyx layer is also an impediment to the transvascular passage of "non-spherical" particles with sizes at the cusp of the physiologic upper limit of pore size, for example, monoclonal antibodies (immunoglobulin G, IgG), which are approximately 11 nm in size based on the calculation of antibody diffusion coefficients in viscous fluids [57]. Therefore, the 12 nm physiologic upper limit of pore size is the likely reason why monoclonal antibody-based systemic chemotherapy has not been effective at treating cancers [58, 59].

## 6. NANOPARTICLE BLOOD HALF-LIFE AND ACCUMULATION WITHIN INDIVIDUAL CANCER CELLS

Spherical particles smaller than 6 nm in diameter (molecular weights less than 40 to 50 kDa) possess relatively short blood half-lives, which is the case of Gd-G1 through standard Gd-G4 dendrimers (Fig. **8**) [13]. These lower generation Gd-dendrimers maintain peak blood concentrations of only minutes, since particle sizes are small enough to be efficiently filtered by the kidney glomeruli. Therefore, lower generation Gd-dendrimers only remain temporarily within the extravascular compartment of cancer tissue (Figs. **4** and **5**) [13], which is not sufficient time for particles to accumulate to therapeutic concentrations within individual cancer cells. Since the blood half-lives of small molecule chemotherapy drugs are even shorter than those of the smallest Gd-dendrimers, the Gd-G1 dendrimers (Figs. **4** and **5**) [13], these small molecule drugs do not accumulate to therapeutic concentrations within individual cancer cells.

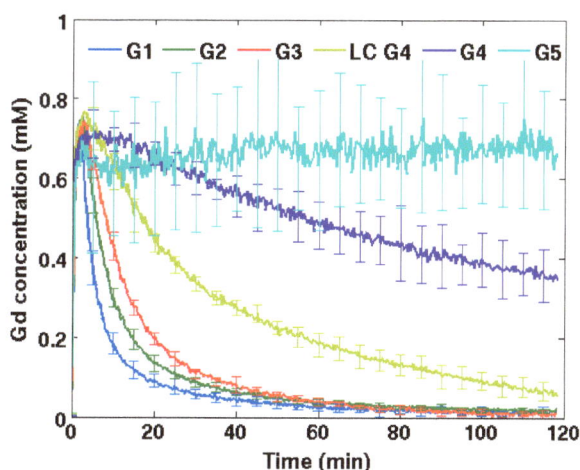

**Figure 8:** Dynamic contrast-enhanced MRI-based blood concentrations of Gd-dendrimer generations over time in rodents.

Gd-G1 (n=4), Gd-G2 (n=6), Gd-G3 (n=6), lowly conjugated (LC) Gd-G4 (n=4), standard Gd-G4 (n=6), Gd-G5 (n=6), Gd-G6 (n=5), Gd-G7 (n=5), and Gd-G8 (n=6). Gd-G6, Gd-G7, and Gd-G8 dendrimer concentration profiles not shown for purposes of figure clarity. Error bars represent standard deviations. Blood Gd concentrations of Gd-dendrimers measured over time in the superior sagittal sinus. Respective Gd-dendrimer generations administered intravenously over 1 minute at a Gd dose of 0.09 mmol Gd/kg animal body weight. Adapted from reference [13].

Spherical particles larger than 7 nm in diameter (molecular weights greater than 70 to 80 kDa) [39, 60, 61] possess relatively long blood half-lives, which is the case of Gd-G5 through Gd-G8 dendrimers (Fig. **8**) [13]. These higher generation Gd-dendrimers maintain peak blood concentrations for several hours, since particle sizes are too large to be efficiently filtered by the kidney glomeruli. Of the higher generation Gd-dendrimers, Gd-G5 and Gd-G6 dendrimers are the two Gd-dendrimer generations that possess long blood half-lives and are also small enough to extravasate across, even, the BTB of small RG-2 brain cancers (Fig. **5**) [13]. Therefore, Gd-G5 and Gd-G6 dendrimers, being within the 7 to 10 nm size range, accumulate to high concentrations within the extravascular compartment of cancer tissue over time (Figs. **4, 5,** and **6**) [12, 13]. Gd-G5 dendrimers remain within the extravascular compartment of cancer tissue for a prolonged period of time after intravenous infusion (Fig. **7**) [12]. This can be visualized on Gd concentration maps from repeated dynamic contrast-enhanced MRI sessions over 12 hours (Fig. **9**).

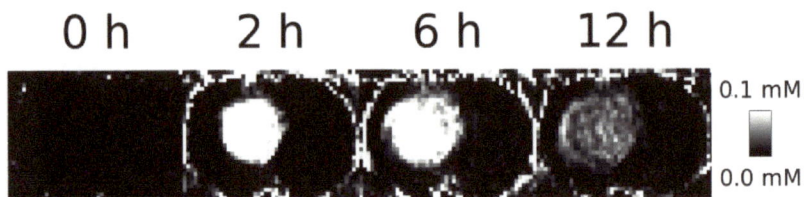

**Figure 9:** $T_1$-weighted dynamic contrast-enhanced MRI-based Gd concentration maps of Gd-G5 dendrimer distribution in the extravascular compartment of RG-2 brain cancer tissue over 12 hours.

Gd-G5 dendrimer administered intravenously over 1 minute at a Gd dose of 0.09 mmol Gd/kg animal body weight at 0 hour time point. Scale from 0 mM [Gd] to 0.1 mM [Gd].

There is significant Gd-G5 dendrimer accumulation within individual RG-2 brain cancer cells, since cancer cells within the extravascular compartment are exposed for several hours to high Gd-G5 dendrimer concentrations. This is apparent on fluorescence microscopy of RG-2 brain cancer tissue harvested 2 hours following the intravenous administration of rhodamine B dye labeled Gd-G5 dendrimers (Fig. **10**, panel A) [13]. Rhodamine B dye labeled Gd-G5 dendrimers accumulate to high concentrations within individual RG-2 brain cancer cells (Fig. **10**, panel A, top right). The lack of rhodamine B dye labeled Gd-G8 dendrimer accumulation within the extravascular compartment of RG-2 brain cancer tissue and within individual RG-2 brain cancer cells is also apparent on fluorescence microscopy (Fig. **10**, panel B) [13].

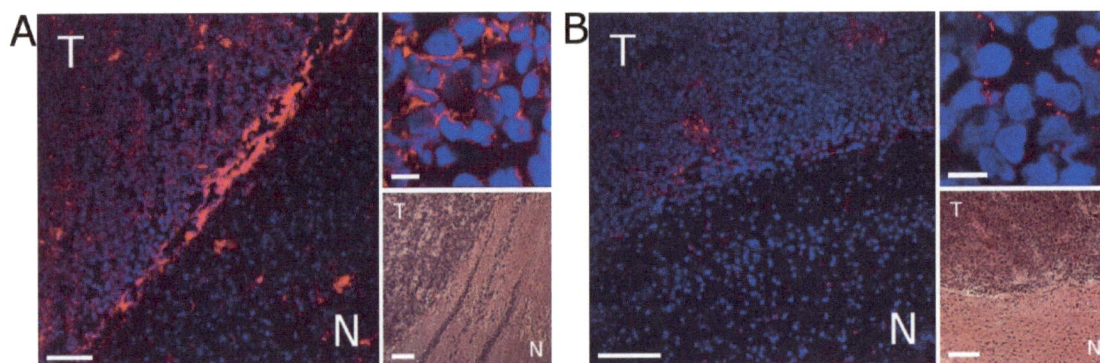

**Figure 10:** *Ex vivo* fluorescence microscopy and histology of RG-2 brain cancer and surrounding normal brain specimens harvested 2 hours following the intravenous bolus of either rhodamine B labeled Gd-G5 dendrimers or rhodamine B labeled Gd-G8 dendrimers.

A) Rhodamine B labeled Gd-G5 dendrimer specimen; RG-2 glioma volume = 31 mm$^3$. B) Rhodamine B labeled Gd-G8 dendrimer specimen; RG-2 glioma volume = 30 mm$^3$. Shown are merged images of blue fluorescence from nuclear stain and red fluorescence from rhodamine B labeled Gd-dendrimers. T = tumor, N = normal; Low power image scale bar = 100 μm; High power image scale bar = 20 μm; H&E image scale bar = 100 μm. Rhodamine B labeled Gd-G5 dendrimers and rhodamine B labeled Gd-G8 dendrimers administered intravenously over 1 minute at a Gd dose of 0.06 mmol Gd/kg animal body weight. Adapted from reference [13].

## 7. SIGNIFICANCE OF POSITIVE CHARGE ON THE NANOPARTICLE SURFACE

Small molecules and peptides with significant focal positive charges are toxic to the luminal glycocalyx layer [62-64], which bears an overall negative charge [65, 66]. Since these positively charged molecules are small molecules that have short blood half-lives, any positive charge-induced toxicity to the glycocalyx would not be readily apparent, except when such positively charged small molecules are administered at high doses [64]. In the case of nanoparticles with long blood half-lives that bear positively charged small molecules on the exterior, the prolonged exposure of the glycocalyx to positive charge results in its disruption over the pathologic BTB [13, 17, 48, 67, 68] as well as over the healthy endothelial barrier, such as the normal blood-brain barrier (BBB) [13, 17].

The positive charge on exterior of naked PAMAM dendrimers is neutralized by the conjugation of Gd-DTPA (charge -2) to a proportion of the surface terminal amine groups. However, when cationic molecules, for example, rhodamine B dye, are conjugated to Gd-dendrimer terminal amine groups, the cationic dye molecules protrude above the conjugated Gd-DTPA moieties and re-introduce positive charge to the particle surface [13, 17]. Therefore, rhodamine B labeled Gd-dendrimers with long blood half-lives disrupt the glycocalyx layer over the already porous BTB of RG-2 brain cancer microvasculature as well as over the normally non-porous BBB of brain microvasculature [13, 17], which is evident *in vivo* on dynamic contrast-enhanced MRI. In the case of rhodamine B labeled Gd-G5 dendrimers, there is profound contrast enhancement of RG-2 brain cancer tissue (Fig. **11**, top row) and contrast enhancement of cancer-free brain tissue, where there normally should not be any contrast enhancement (Fig. **11**, top row arrow). In the case of rhodamine B labeled Gd-G8 dendrimers, there is some contrast enhancement of RG-2 brain cancer tissue (Fig. **11**, bottom row arrow), which does not occur with the Gd-G8 dendrimers.

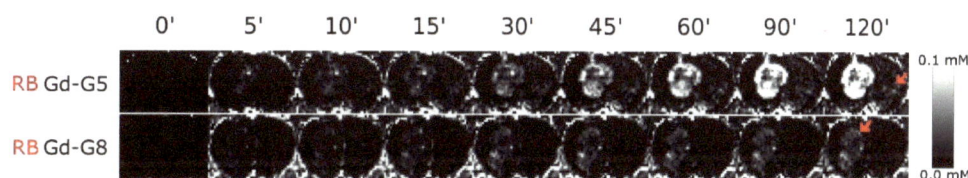

**Figure 11:** Representative $T_1$-weighted dynamic contrast-enhanced MRI-based Gd concentration maps of rhodamine B labeled Gd-G5 (RB Gd-G5) and rhodamine B labeled Gd-G8 (RB Gd-G8) dendrimer distribution in RG-2 brain cancer and normal brain over time.

Top red arrow highlights non-selective contrast enhancement in normal brain tissue, and bottom arrow highlights contrast enhancement in RG-2 brain cancer tissue. Rhodamine B labeled Gd-G5 and rhodamine B labeled Gd-G8 dendrimers administered intravenously over 1 minute at a Gd dose of 0.06 mmol Gd/kg animal body weight. Scale from 0 mM [Gd] to 0.1 mM [Gd].

The *in vivo* dynamic contrast-enhanced MRI findings are consistent with what is observed *ex vivo* on fluorescence microscopy of harvested RG-2 brain cancer tissue and surrounding normal brain tissue, as there is some accumulation of rhodamine B labeled Gd-G5 dendrimers in cancer-free surrounding normal brain tissue (Fig. **10**, panel A) [13] and also some accumulation of rhodamine B labeled Gd-G8 dendrimers in RG-2 brain cancer tissue (Fig. **10**, panel B) [13]. Based on these findings, the enhanced transvascular extravasation of cationic nanoparticles across the BTB and the non-selective transvascular extravasation of these particles across normal endothelial barriers is by positive charge-induced toxicity to the luminal glycocalyx layer.

## 8. NUCLEAR PORE SIZE LIMITATION TO NANOPARTICLE ENTRY INTO CANCER CELL NUCLEI

Nanoparticles within the 7 to 10 nm size range can extravasate across the BTB of cancer microvasculature and accumulate to high concentrations within individual cancer cells. Therefore, such nanoparticles bearing chemotherapy can be used to deliver therapeutic concentrations of small molecule chemotherapy drugs into individual cancer cells. Since classical small molecule chemotherapy drugs act at the level of the cell nucleus and interfere with DNA replication, for these drugs to be effective, drug fraction must accumulate to high concentrations within cancer cell nuclei. For this reason, the nuclear pore size limitation to nanoparticle entry into cancer cell nuclei needs to be taken into consideration [13, 17].

The nuclear pore size limitation to nanoparticle entry into cancer cell nuclei is evident on *in vitro* fluorescence microscopy of RG-2 glioma cells incubated in media containing rhodamine B labeled Gd-G2, rhodamine B labeled Gd-G5, or rhodamine B labeled Gd-G8 dendrimers (Fig. **12**) [13]. All of these rhodamine labeled Gd-dendrimer generations accumulate in the cell cytoplasm (Fig. **12**, all panels), however, only the rhodamine labeled Gd-G2 dendrimers are small enough to transverse across the nuclear pores and accumulate within the cancer cell nuclei (Fig. **12**, left panel).

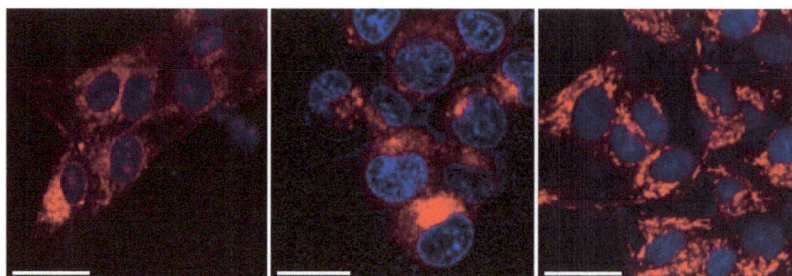

**Figure 12:** *In vitro* fluorescence microscopy of cultured RG-2 glioma cells incubated for 4 hours in media containing rhodamine B labeled Gd-G2 dendrimers (left), rhodamine B labeled Gd-G5 dendrimers (middle), or rhodamine B labeled Gd-G8 dendrimers (right) at a concentration of 7.2 μM with respect to rhodamine B.

Shown are merged images of blue fluorescence from nuclear stain and red fluorescence from rhodamine B labeled Gd-dendrimers. Scale bars = 20 μm. Adapted from reference [13].

Since rhodamine B labeled Gd-G5 dendrimers are larger than the nuclear pore size in cancer cell nuclei, other nanoparticles within the 7 to 10 nm size range would also be too large to enter cancer cell nuclei. Therefore, for nanoparticles within the 7 to 10 nm size range to function as effective drug delivery devices, drug fraction must be released into the cytoplasm following nanoparticle endocytosis, as the released drug fraction could accumulate to high concentrations within cancer cell nuclei [13, 17].

## 9. SIGNIFICANCE OF THE PH-SENSITIVE HYDRAZONE BOND FOR DRUG ATTACHMENT AND RELEASE

The pH-sensitive hydrazone bond is a covalent bond that is stable at the physiologic pH of 7.4, which is the approximate pH of blood, and labile at the acidic pH of 5.5, which is the approximate pH in cellular lysosomal compartments [33-35, 69]. Due to the stability of the hydrazone bond at physiologic pH, there is minimal free drug release from nanoparticles in systemic circulation, prior to nanoparticle extravasation across the BTB, and as a result, little systemic toxicity from circulating nanoparticles bearing chemotherapy. There is also minimal free drug release within the extravascular extracellular compartment of cancer tissue, after nanoparticle extravasation across the BTB, since the extravascular extracellular compartment of cancer tissue is significantly less acidotic than intracellular lysosomal compartments [70]. However, on nanoparticle endocytosis into the acidic environment of cancer cell lysosomal compartments, the hydrazone bond hydrolyzes relatively quickly and there is efficient intracellular drug release in the cell cytoplasm, which results in permeation of free drug fraction across nuclear pores and drug accumulation in cancer cell nuclei [17]. Since nanoparticles within the 7 to 10 nm size range are too large to enter cancer cell nuclei [13], the pH-sensitive hydrazone bond is a useful covalent bond for the attachment of chemotherapy drugs, as it enables efficient drug release following nanoparticle endocytosis, which allows for drug fraction to accumulate to effective concentrations in cancer cell nuclei [17].

## 10. INITIAL THERAPEUTIC EFFICACY OF THE GD-G5-DOXORUBICIN DENDRIMER IN THE RG-2 BRAIN CANCER MODEL

The Gd-G5-doxorubicin dendrimer is an imageable nanoparticle bearing chemotherapy within the 7 to 10 nm size range with doxorubicin conjugated to the terminal amines *via* a hydrazone bond [17]. Since the Gd-G5-doxorubicn dendrimer is imageable, the extent of Gd-G5-doxorubicin dendrimer accumulation in the extravascular compartment of cancer tissue is a surrogate measure of the extent of doxorubicin accumulation in the extravascular compartment. The Gd-G5-doxorubicin dendrimer accumulates to high concentrations in the extravascular compartment of RG-2 brain cancer tissue within 60 minutes of intravenous infusion (Fig. **13**) [17]. There is some non-selective accumulation of the Gd-G5-doxorubicin dendrimer in normal brain tissue (Fig. **13**, arrow), which is indicative of positive charge-induced toxicity to the glycocalyx of the normal BBB from the presence of doxorubicin, a cationic drug, on the dendrimer surface [64].

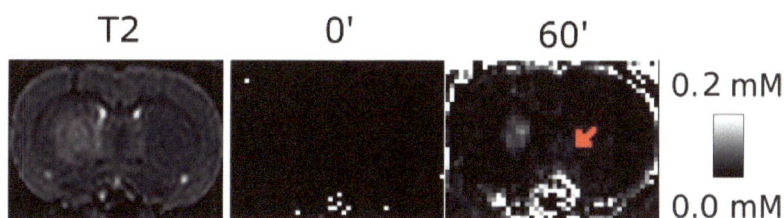

**Figure 13:** $T_2$-weighted anatomic MRI scan image of RG-2 brain cancer and $T_1$-weighted dynamic contrast-enhanced MRI scan Gd concentration maps at baseline and at 60 minutes following Gd-G5-doxorubicn dendrimer infusion.

Red arrow highlights non-selective contrast enhancement in normal brain tissue. The Gd-G5-doxorubicin dendrimer administered intravenously over 2 minutes at a Gd dose of 0.09 mmol Gd/kg animal body weight, which is equivalent to a doxorubicin dose of 8 mg/kg. Scale from 0 mM [Gd] to 0.2 mM [Gd]. Adapted from reference [17].

In RG-2 brain cancer regression studies, the Gd-G5-doxorubicin dendrimer has demonstrated therapeutic efficacy [17], as one 8 mg/kg dose of Gd-G5-doxorubicin dendrimer with respect to doxorubicin is significantly more effective than one 8 mg/kg dose of free doxorubicin at inhibiting the growth of RG-2 brain cancers for 24 hours (Fig. **14**) [17].

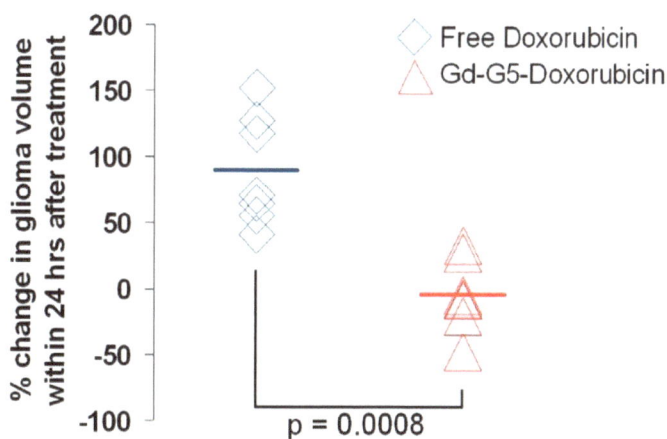

**Figure 14:** Percent change in RG-2 brain cancer volume within 24 hours after one intravenous 8 mg/kg dose of Gd-G5-doxorubicin dendrimer with respect to doxorubicin (n=7) or 8 mg/kg dose of free doxorubicin (n=7).

Percent change in RG-2 glioma volumes were calculated based on whole tumor volumes measured on initial $T_2$-weighted MRI scans acquired at the time of agent infusion for both groups, and on repeat $T_2$-weighted MRI scans within 22 ± 2 hours for the Gd-G5-doxorubicin group and 24 ± 1 hour for the free doxorubicin group. Student's two-tailed paired t-test p value < 0.0008. Adapted from reference [17].

## 11. FUTURE PERSPECTIVE

The Gd-G5-doxorubicin dendrimer is the prototype of an imageable nanoparticle within the 7 to 10 nm size range [17]. The Gd-G5-doxorubicin dendrimer will need to be optimized further, since there is positive charge on the Gd-G5-doxorubicin dendrimer surface due to the presence of doxorubicin that protrudes above the Gd-DTPA moieties on the dendrimer exterior. This results in positive charge-induced toxicity to the glycocalyx of the normal BBB and the non-selective accumulation of the Gd-G5-doxorubicin dendrimer in normal brain tissue to some extent [17]. Due to this drawback of the Gd-G5-doxorubicin dendrimer, in the future, cationic chemotherapy drugs will need to be conjugated to imageable dendrimer terminal groups *via* hydrazone bonds that are closer to the particle interior, as this will limit the introduction of positive charge on the particle exterior. Furthermore, it will also be advantageous to use naked half generation PAMAM dendrimers (*i.e.* G5.5) as substrates for the conjugation of cationic dyes and drugs, as half generation PAMAM dendrimers are anionic dendrimers. Other types of biocompatible dendrimers, for example, those that are amino acid-based, would also be appropriate substrates for dye and drug attachment, as long as there is no net positive charge on the functionalized particle surface.

In summary, spherical nanoparticles in the 7 to 10 nm size range maintain peak blood concentrations for several hours and are sufficiently smaller than the 12 nm physiologic upper limit of pore size in the BTB to accumulate to high concentrations within individual cancer cells [12, 13, 17]. Chemotherapy drug attachment to nanoparticles within this size range *via* the pH-sensitive hydrazone bond results in efficient intracellular drug release and drug accumulation in cancer cell nuclei. The initial *in vivo* therapeutic efficacy of the Gd-G5-doxorubicin dendrimer in RG-2 brain cancer regression studies stems from the effective transvascular delivery of doxorubicin across the BTB of brain cancer microvasculature into individual cancer cells and doxorubicin accumulation to therapeutic concentrations within cancer cell nuclei [17]. Therefore, it is foreseeable that such imageable nanoparticles bearing chemotherapy, which are within the 7 to 10 nm size range, will also demonstrate therapeutic efficacy in the treatment of cancers located outside the brain and central nervous system.

## ACKNOWLEDGEMENTS

This research was funded by the National Institute of Biomedical Imaging and Bioengineering, and the Clinical Center Radiology and Imaging Sciences Program.

## REFERENCES

[1]    Folkman J. Tumor angiogenesis: therapeutic implications. N Engl J of Med 1971; 285(21): 1182-1186.
[2]    Folkman J, Klagsbrun M. Angiogenic factors. Science 1987; 235(4787): 442-447.
[3]    Senger DR, Perruzzi CA, Feder J, *et al.* A highly conserved vascular permeability factor secreted by a variety of human and rodent tumor cell lines. Cancer Res 1986; 46(11): 5629-32.
[4]    Roberts WG, Palade GE. Neovasculature induced by vascular endothelial growth factor is fenestrated. Cancer Res 1997; 57(4): 765-772.
[5]    Jain RK. Transport of molecules across tumor vasculature. Cancer Metastasis Rev 1987; 6(4): 559-593.
[6]    Michel CC, Curry FE. Microvascular permeability. Physiol Rev 1999; 79(3): 703-761.
[7]    Vargas F, Johnson JA. An estimate of reflection coefficients for rabbit heart capillaries. J Gen Physiol 1964; 47: 667-677.
[8]    Fenstermacher JD, Johnson JA. Filtration and reflection coefficients of the rabbit blood-brain barrier. Am J Physiol 1966; 211(2): 341.
[9]    Bruns RR, Palade GE. Studies on blood capillaries. I. General organization of blood capillaries in muscle. J Cell Biol 1968; 37(2): 244-276.
[10]   Satchell SC, Braet F. Glomerular endothelial cell fenestrations: an integral component of the glomerular filtration barrier. Am J Physiol Renal Physiol 2009; 296(5): F947-956.
[11]   Sarin H, Kanevsky AS, Fung SH, *et al.* Metabolically stable bradykinin B2 receptor agonists enhance transvascular drug delivery into malignant brain tumors by increasing drug half-life. J Transl Med 2009; 7: 33.
[12]   Sarin H, Kanevsky AS, Wu H, *et al.* Physiologic upper limit of pore size in the blood-tumor barrier of malignant solid tumors. J of Transl Med 2009; 7: 51.
[13]   Sarin H, Kanevsky AS, Wu H, *et al.* Effective transvascular delivery of nanoparticles across the blood-brain tumor barrier into malignant glioma cells. J of Transl Med 2008; 6: 80.
[14]   Vick NA, Bigner DD. Microvascular abnormalities in virally-induced canine brain tumors. Structural bases for altered blood-brain barrier function. J Neurol Sci 1972; 17(1): 29-39.
[15]   Hashizume H, Baluk P, Morikawa S, *et al.* Openings between defective endothelial cells explain tumor vessel leakiness. Am J Pathol 2000; 156(4): 1363-1380.
[16]   Schlageter KE, Molnar P, Lapin GD, *et al.* Microvessel organization and structure in experimental brain tumors: Microvessel populations with distinctive structural and functional properties. Microvasc Res 1999; 58(3): 312-328.

[17]  Sarin H. Recent progress towards development of effective systemic chemotherapy for the treatment of malignant brain tumors. J Transl Med 2009; 7: 77.

[18]  Vick NA, Khandekar JD, Bigner DD. Chemotherapy of brain tumors. The "blood-brain barrier" is not a factor. Arch Neurol 1977; 34(9): 523-526.

[19]  Groothuis DR, Vick NA. Brain tumors and the blood-brain barrier. Trends Neurosci 1982; 5: 232-235.

[20]  Tannock IF, Lee CM, Tunggal JK, *et al.* Limited penetration of anticancer drugs through tumor tissue: A potential cause of resistance of solid tumors to chemotherapy. Clin Cancer Res 2002; 8(3): 878-884.

[21]  Norton L. Evolving concepts in the systemic drug therapy of breast cancer. Semin Oncol 1997; 24(4 Suppl 10): 3-10.

[22]  Denny WA. DNA-intercalating ligands as anti-cancer drugs: prospects for future design. Anticancer Drug Des 1989; 4(4): 241-263.

[23]  Shawver LK, Slamon D, Ullrich A. Smart drugs: Tyrosine kinase inhibitors in cancer therapy. Cancer Cell 2002; 1(2): 117-123.

[24]  Early Breast Cancer Trialists' Collaborative Group (EBCTCG). Effects of chemotherapy and hormonal therapy for early breast cancer on recurrence and 15-year survival: An overview of the randomised trials. Lancet 2005; 365(9472): 1687-1717.

[25]  Ahmann DL, Creagan ET, Hahn RG, *et al.* Complete responses and long-term survivals after systemic chemotherapy for patients with advanced malignant melanoma. Cancer 1989; 63(2): 224.

[26]  Burroughs A, Hochhauser D, Meyer T. Systemic treatment and liver transplantation for hepatocellular carcinoma: Two ends of the therapeutic spectrum. Lancet Oncol 2004; 5(7): 409-418.

[27]  Wyman K, Atkins MB, Prieto V, *et al.* Multicenter phase II trial of high-dose imatinib mesylate in metastatic melanoma: Significant toxicity with no clinical efficacy. Cancer 2006; 106(9): 2005-2011.

[28]  Soo Choi H, Liu W, Misra P, *et al.* Renal clearance of quantum dots. Nat Biotechnol 2007; 25(10): 1165-1170.

[29]  Langer R. Drug delivery and targeting. Nature 1998; 392(6679 Suppl): 5-10.

[30]  Maeda H, Greish K, Fang J. The EPR effect and polymeric drugs: A paradigm shift for cancer chemotherapy in the 21$^{st}$ century. Adv Polymer Sci 2006; 193: 103-121.

[31]  Lengyel JS, Milne JLS, Subramaniam S. Electron tomography in nanoparticle imaging and analysis. Nanomedicine (Lond) 2008; 3(1): 125-131.

[32]  Bootz A, Vogel V, Schubert D, *et al.* Comparison of scanning electron microscopy, dynamic light scattering and analytical ultracentrifugation for the sizing of poly(butyl cyanoacrylate) nanoparticles. Eur J Pharm Biopharm 2004; 57(2): 369-375.

[33]  Lee CC, Gillies ER, Fox ME, *et al.* A single dose of doxorubicin-functionalized bow-tie dendrimer cures mice bearing C-26 colon carcinomas. Proc Natl Acad Sci USA 2006; 103(45): 16649-16654.

[34]  Greenfield RS, Kaneko T, Daues A, *et al.* Evaluation *in vitro* of adriamycin immunoconjugates synthesized using an acid-sensitive hydrazone linker. Cancer Res 1990; 50(20): 6600-6607.

[35]  Kono K, Kojima C, Hayashi N, *et al.* Preparation and cytotoxic activity of poly(ethylene glycol)-modified poly(amidoamine) dendrimers bearing adriamycin. Biomaterials 2008; 29(11): 1664-1675.

[36]  Tomalia DA, Frechet JM. Discovery of dendrimers and dendritic polymers: a brief historical perspective. J Polym Sci Part A: Polym Chem 2002; 40(16): 2719-2728.

[37]  Tomalia DA, Reyna LA, Svenson S. Dendrimers as multi-purpose nanodevices for oncology drug delivery and diagnostic imaging. Biochem Soc Trans 2007; 35(Part 1): 61-67.

[38]  Kobayashi H, Brechbiel MW. Nano-sized MRI contrast agents with dendrimer cores. Adv Drug Deliv Rev 2005; 57(15): 2271-2286.

[39]  Sousa AA, Aronova MA, Wu H, *et al.* Determining molecular mass distributions and compositions of functionalized dendrimer nanoparticles. Nanomedicine 2009; 4(4): 391-399.

[40]  Knopp MV, Weiss E, Sinn HP, *et al.* Pathophysiologic basis of contrast enhancement in breast tumors. J Magn Reson Imaging 1999; 10(3): 260-266.

[41]  Tofts PS, Kermode AG. Measurement of the blood-brain barrier permeability and leakage space using dynamic MR imaging. 1. Fundamental concepts. Magn Reson Med 1991; 17(2): 357-367.

[42]  Tofts PS, Berkowitz B, Schnall MD. Quantitative analysis of dynamic Gd-DTPA enhancement in breast tumors using a permeability model. Magn Reson Med 1995; 33(4): 564-568.

[43]  Caravan P, Farrar CT, Frullano L, *et al.* Influence of molecular parameters and increasing magnetic field strength on relaxivity of gadolinium- and manganese-based T1 contrast agents. Contrast Media Mol Imaging 2009; 4(2): 89-100.

[44] Haacke EM, Brown RW, Thompson MR, *et al.* Magnetic Resonance Imaging: Physical Principles and Sequence Design. New York: Wiley; 1999.

[45] Aas AT, Brun A, Blennow C, *et al.* The RG2 rat glioma model. J Neurooncol 1995; 23(3): 175-183.

[46] Barth RF. Rat brain tumor models in experimental neuro-oncology: The 9L, C6, T9, F98, RG2 (D74), RT-2 and CNS-1 gliomas. J Neurooncol 1998.; 36(1): 91-102.

[47] Roberts WG, Delaat J, Nagane M, *et al.* Host microvasculature influence on tumor vascular morphology and endothelial gene expression. Am J Pathol 1998; 153(4): 1239-1248.

[48] Hobbs SK, Monsky WL, Yuan F, *et al.* Regulation of transport pathways in tumor vessels: role of tumor type and microenvironment. Proc Natl Acad Sci USA 1998; 95(8): 4607-4612.

[49] Monsky WL, Carreira CM, Tsuzuki Y, *et al.* Role of host microenvironment in angiogenesis and microvascular functions in human breast cancer xenografts: Mammary fat pad versus cranial tumors. Clin Cancer Res 2002; 8(4): 1008-1013.

[50] Curry FE, Michel CC. A fiber matrix model of capillary permeability. Microvasc Res 1980; 20(1): 96-99.

[51] Squire JM, Chew M, Nneji G, *et al.* Quasi-periodic substructure in the microvessel endothelial glycocalyx: a possible explanation for molecular filtering? J Struct Biol 2001; 136(3): 239-255.

[52] Chappell D, Jacob M, Paul O, *et al.* The glycocalyx of the human umbilical vein endothelial cell: an impressive structure *ex vivo* but not in culture. Circ Res 2009; 104(11): 1313-1317.

[53] Muldoon LL, Nilaver G, Kroll RA, *et al.* Comparison of intracerebral inoculation and osmotic blood-brain barrier disruption for delivery of adenovirus, herpesvirus, and iron oxide particles to normal rat brain. Am J Pathol 1995; 147(6): 1840-1851.

[54] Nilaver G, Muldoon LL, Kroll RA, *et al.* Delivery of herpesvirus and adenovirus to nude rat intracerebral tumors after osmotic blood-brain barrier disruption. Proc Natl Acad Sci USA 1995; 92(21): 9829-9833.

[55] Lossinsky AS, Shivers RR. Structural pathways for macromolecular and cellular transport across the blood-brain barrier during inflammatory conditions. Rev Histol Histopathol 2004; 19(2): 535-564.

[56] Rapoport SI. Osmotic opening of the blood-brain barrier: principles, mechanism, and therapeutic applications. Cell Mol Neurobiol 2000; 20(2): 217-230.

[57] Saltzman WM, Radomsky ML, Whaley KJ, *et al.* Antibody diffusion in human cervical mucus. Biophys J 1994; 66(2 Part I): 508-515.

[58] Kalofonos HP, Grivas PD. Monoclonal antibodies in the management of solid tumors. Curr Top Med Chem 2006; 6(16): 1687-1705.

[59] Yang W, Wu G, Barth RF, *et al.* Molecular targeting and treatment of composite EGFR and EGFRvIII-positive gliomas using boronated monoclonal antibodies. Clin Cancer Res 2008; 14(3): 883-891.

[60] Li J, Piehler LT, Qin D, *et al.* Visualization and characterization of poly(amidoamine) dendrimers by atomic force microscopy. Langmuir 2000; 16(13): 5613-5616.

[61] Müller R, Laschober C, Szymanski WW, *et al.* Determination of molecular weight, particle size, and density of high number generation pamam dendrimers using maldi-tof-ms and nes-gemma. Macromolecules 2007; 40(15): 5599-5605.

[62] Lutty GA. The acute intravenous toxicity of biological stains, dyes, and other fluorescent substances. Toxicol Appl Pharmacol 1978; 44(2): 225-249.

[63] Herce HD, Garcia AE. Cell penetrating peptides: How Do They Do It? J Biol Phys 2008; 33: 345-356.

[64] Jeansson M, Bjorck K, Tenstad O, *et al.* Adriamycin alters glomerular endothelium to induce proteinuria. J Am Soc Nephrol 2009; 20(1): 114-122.

[65] Pries AR, Secomb TW, Gaehtgens P. The endothelial surface layer. Pflugers Arch 2000; 440(5): 653-666.

[66] Hardebo JE, Kahrstrom J. Endothelial negative surface charge areas and blood-brain barrier function. Acta Physiol Scand 1985; 125(3): 495-499.

[67] Dellian M, Yuan F, Trubetskoy VS, *et al.* Vascular permeability in a human tumour xenograft: Molecular charge dependence. Br J Cancer 2000; 82(9): 1513-1518.

[68] Campbell RB, Fukumura D, Brown EB, *et al.* Cationic charge determines the distribution of liposomes between the vascular and extravascular compartments of tumors. Cancer Res 2002; 62(23): 6831-6836.

[69] Moriyama Y, Maeda M, Futai M. Involvement of a non-proton pump factor (possibly Donnan-type equilibrium) in maintenance of an acidic pH in lysosomes. FEBS Lett 1992; 302(1): 18-20.

[70] Vaupel P. Blood flow and metabolic microenvironment of brain tumors. J Neurooncol 1994; 22(3): 261-267.

# Nanocrystalline Silver: Use in Wound Care

## Valerie Edwards-Jones[*]

*Manchester Metropolitan University, UK*

**Abstract:** The antimicrobial activity of nanocrystalline metals has been shown to be more sustained than normal metal ions. This chapter reviews the use of nanocrystalline silver in wound care and the impact these dressings are having on acute and chronic wounds in terms of healing, toxicity and reduction of biological burden. Production of the nanocrystalline dressings will be described and the antimicrobial mode of action of silver described. The *in-vitro* and *in-vivo* assessment of nanocrystalline silver shows these dressings to be advantageous for the health care practioner and an excellent addition to wound care productions.

**Keywords:** Nanocrystalline metals; wound care; wound healing; toxicity; antimicrobial; silver; nanoparticles; immunomodulation; anti-microbial; infectyion; burns.

## BACKGROUND

Silver has been used since antiquity and has been described for its antimicrobial properties on numerous occasions. Metallic silver is not soluble in water and as such is not antimicrobial although there may be some minor leeching of silver ions in the presence of water (hence why there were silver coins in a fountain and the Romans used to keep water fresh in silver urns). The first documented use of silver in modern wound care was silver nitrate soaked to prevent infection in burns [1, 2]. These required up to four dressing changes daily to ensure they worked effectively and negated some of the positive effects in terms of wound healing. Moyer demonstrated that these soaks were effective against typical wound pathogens *e.g.* *Staphylococcus aureus*, *Pseudomonas aeruginosa* and haemolytic *Streptococci*.

In the early 1960s, silver sulphadiazine (SSD), a slow release silver compound produced by a chemical reaction between silver nitrate and sodium sulphadiazine was produced to reduce the number of dressing changes [3]. This still remains an effective topical prophylactic agent and treatment for a number of different infected wounds. SSD is relatively insoluble and releases the silver ions into the wound bed slowly over a twenty four hour period.

Since the introduction of silver as an antimicrobial agent in wound care a variety of compounds have been incorporated into dressings for the prevention and treatment of wound infection. Modern silver dressings are very sophisticated and offer benefits of physical protection, prevention of infection, reducing odour, absorbing excess exudates and maintaining a moist environment at the wound surface [4]. Dressings vary from films, foams, fibrous products, beads, hydrogels or hydrocolloid dressings. The costs of the dressings vary considerably and the cost for silver dressings in the UK has increased from £800K in 2004 to £23million in 2006 [5]. Silver dressings are commercially available in several forms and differ on levels and release of silver. Metallic silver is insoluble and therefore inactive as an antimicrobial. However, it is chemically reactive and forms a variety of silver salts. Following dissociation of these salts mediated through moisture, the active ionic state is formed. Four different states of silver exist; $Ag^+$, $Ag^{++}$, $Ag^{+++}$ and $Ag0$. Singly charged silver, $Ag^+$, is the most biologically active and is dependent upon solubility [6]. $Ag^{++}$ and $Ag^{+++}$ show some activity but are more likely to form insoluble complexes and be rapidly inactivated by protein binding [7].

## 1. NANOCRYSTALLINE SILVER IN WOUND DRESSINGS

Nanocrystals or $Ag^+$ nanoparticles can be produced using a variety of methods including vapour deposition

*Address correspondence to Valerie Edwards-Jones: Manchester Metropolitan University, Ormond Building, Lower Ormond Street, Manchester, M15 6BX, UK; Tel: ++44 (0)161 247 1025; Fax: ++44 (0)161 247 6823; E-mail: v.e.jones@mmu.ac.uk

**Mark Slevin (Ed)**

[8], chemical reduction [9] and laser ablation [10]. The physical, biochemical and antimicrobial properties of different size and shapes of these $Ag^+$ nanoparticles vary and also portray different optical, magnetic and catalytic properties. The particle size of $Ag^+$ nanoparticles produced by laser ablation can be controlled [10] and improved antibacterial activity is dependent on both the size and the shape of the Ag nanoparticles [9]. The nanocrystalline dressings in common use is Acticoat [TM]; Smith and Nephew Healthcare and is produced by surface vapour deposition of nanocrystalline silver particles onto the surface of the wound dressing.

## 2. METHODS OF PRODUCTION OF NANOPARTICLES OF SILVER

### 2.1 Vapour Deposition Method

The vapour deposition methods involves passing an electric current through Argon gas in a vacuum chamber (which acts as an anode) containing a silver cathode. The silver is displaced from the cathode following impact of the positively charged argon ions and are then deposited as Ag nanocrystals on the substrate to be coated when energy inputs are limited (see Fig. 1). The nanocrystals are approximately 20-120 nanometres in size when sputtered by this method onto a surface and create a larger surface area and availability of Ag ions. It has been demonstrated that there is a continual sustained release of $Ag^+$ when exposed to water, occurring over 3-7 days, the levels remaining at approximately 100 parts per million [8].

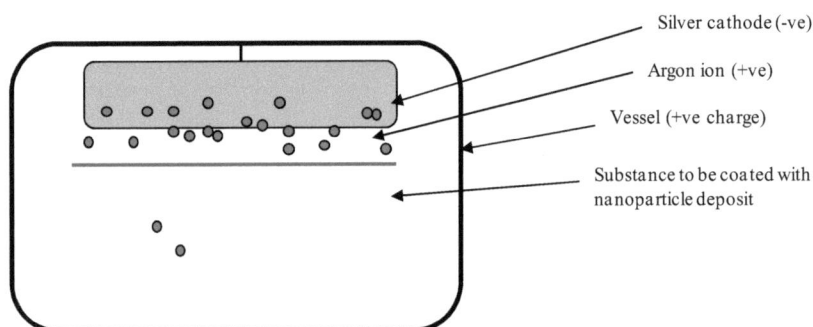

Figure 1: Shows the diagrammatic representation of the vapour deposition method. The vessel is under vacuum and is positively charged. Temperature is low and pressure is high.

### 2.2 Chemical Reduction Method

Nanoparticles can also be produced by Lee and Meisel's method as described by Pal *et al*, and Zhao *et al* [9, 10]. Briefly this involves adding trisodium citrate to boiling silver nitrate solution (with agitation) until a green-gray silver colloid is obtained. Following purification the nanoparticles are freeze dried for future use.

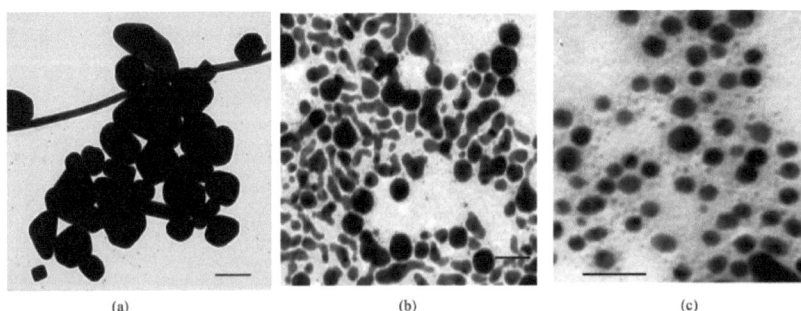

Figure 2: TEM images of silver colloids particles prepared with (a) chemical reduction, (b) laser ablation with 532 nm and (c) with 248 nm.

### 2.3 The Laser Ablation Method

The Laser ablation method uses nitric acid cleaned and sonic washed silver foil (1mm thickness) immersed in a deionized water filled quartz cell as described by [10]. Briefly a Nd:YAG-pulsed laser at 2 wavelengths

(532 nm and 248nm) was used to produce the nanoparticles which are found to be different sizes and shapes dependent upon the ablation method used as shown below in Fig. **2**. These nanoparticles had differing spectrophotometric properties

## 2.4 Solution Phase Method of Preparation of Silver Nanoparticles

Different shapes of Ag nanoparticles, rod shaped and triangular, can be made by a solution phase method as described in [9]. Following production of the nanoparticles by chemical reaction, addition of molar sodium hydroxide accelerates particle growth. The particles are then separated and purified using differential centrifugation at 2,100g for 10 min, 755g for 10 min and 84g for 10 min. The supernatant solution contains the truncated triangular silver nanoparticles (Fig. **3**).

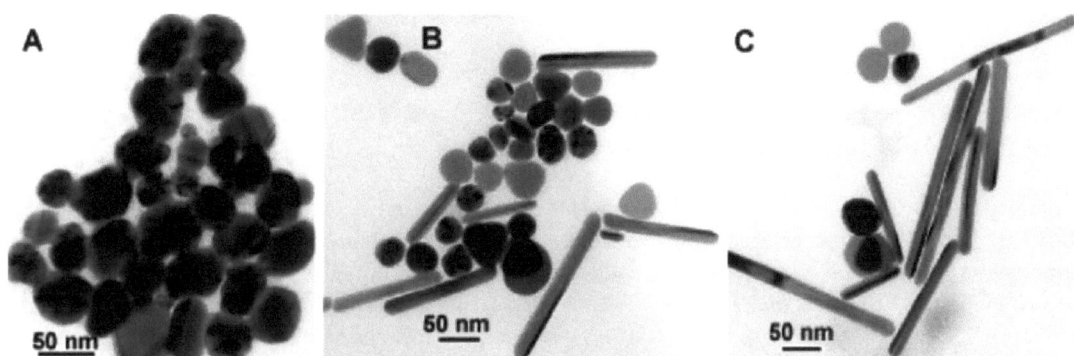

**Figure 3:** EFTEM images of silver nanoparticles. (A) Spherical nanoparticles synthesized by citrate reduction. (B) Silver nanoparticles of different shapes. (C) Purified rod-shaped nanoparticles

## 3. MODE OF ACTION OF SILVER

$Ag^+$ has a broad spectrum of activity and inhibits growth of bacteria and yeasts between 8-80 ppm whilst Gram negative bacteria *e.g. P. aeruginosa* have a lower minimal inhibitory concentration (MIC) than Gram positive bacteria *e.g. S. aureus* [11]. The activity can vary depending upon the conditions used to assess silver activity *in-vitro* and depends upon the availability of silver. Silver acts on multiple sites within the bacterial cell acting on the cell membrane[12, 13] respiratory enzymes and intracellular enzymes by reacting with sulphydral groups [14, 15], and interchelates with DNA binding specifically with GC groups[16, 17, 18]. Silver is a broad spectrum antimicrobial agent showing activity against bacteria, fungi and viruses [19]. Like many antiseptics, silver is quickly inactivated by proteins, phosphates, sulphates and chlorides all frequently found in tissues and necessitate frequent dressing changes unless there is a novel slow release mechanism. Availability of silver ions is therefore important if the dressing is to be used for prolonged periods. A number of studies have reported levels of silver released from the plethora of available dressings and these vary from 1p.p.m to 100 ppm. The levels do vary considerably depending upon the solvent used during *in-vitro* testing. However, it must be recognized that levels of available silver will be different in a wound because of the protein levels and protein binding. The different forms are used in a wide range of products and solubility; ultimately releasing $Ag^+$ determines the longevity of antimicrobial activity. Laboratory studies have shown that release of silver ions from nanocrystalline dressings is sustained and above the minimal inhibitory concentration for up to seven days.

## 4. IMMUNOMODULATORY ACTION OF SILVER

In 2002, Wright *et al.* postulated that nanocrystalline $Ag^+$ may promote wound healing and described the inhibition of matrix metalloproteases (MMP's) known to prevent wound healing and also MMPs produced by *S. aureus* the commonest wound pathogen. In addition, nanocrystalline $Ag^+$ was shown to bind to metallothioneins which contribute to tissue repair and beneficially modulate the immune system [20]. Toxic epidermal necrolysis and Steven Johnson Syndrome showed marked improvement when treated with a topical nanocrystalline $Ag^+$ dressing compared to traditional Vaseline gauze [21, 22]. In animal studies, a topical antimicrobial cream containing 1% w/w nanocrystalline $Ag^+$ suppressed two pro-inflammatory cytokines, IL-12 and TNF-α, known to be involved in allergic skin diseases such as contact dermatitis [23-25]. Inhibition of a number of immunomodulatory factors by nanocrystalline $Ag^+$ in the wound bed could promote wound healing. Therefore silver dressings may be bi-functional depending upon the concentration released into the wound bed, with high levels exerting antimicrobial effects and sub-lethal levels inhibiting MMP's and other pro-inflammatory factors [26].

**Nanocrystalline** Silver dressings can be used as a topical prophylactic agent when there is a high risk of infection, for example in major trauma *e.g.* large burns, in the early stages of wound healing. They can be used in wounds to reduce bacterial biological burden and help facilitate healing and also in wounds that show signs of infection. They should be discontinued once wound biological burden is controlled and wound healing progresses. If no improvement in wound status is noted after 2 weeks of use then they should be changed and alternative dressings tried.

## 5.  *IN-VITRO* AND *IN-VIVO* ASSESSMENT OF ANTIMICROBIAL ACTIVITY OF NANOCRYSTALLINE SILVER

Laboratory studies are essential in the development of any medical device and include a series of tests validating the physical characteristics of the dressing alongside other beneficial effects involved with wound healing. Vital tests such as stability and its pharmacological kinetics need to be evaluated prior to going to markets. Often a second stage involves cultures of tissue cells and finally animal studies. These studies provide the evidence required to determine which products are suitable to take into clinical trials and assess their effects in humans.

Testing the antimicrobial activity of silver eluted or within dressings is tested using the Minimal Inhibitory Concentration (MIC), minimal bacteriocidal concentration (MBC), challenge test, transmission test [27], time-kill assay, zone of inhibition [28], log reduction, [27], and barrier [29, 30]. Although these methods are well described and standardised, there are limitations especially when testing silver as the system has to kept moist to allow continued activity of silver ions. During incubation of foam dressings, the system can dry out and give false negative results. Dressings with high levels of silver will kill the microorganisms rapidly (within 30 minutes or 2 hrs) or show a 3 log reduction. Dressings with low levels of silver do not effect the growth of the microorganisms in the test system. The Zone Of Inhibition (ZOI) test is of limited use as silver does not diffuse readily through the growth medium. Therefore results can be attained where a dressing has an effective time-kill or challenge test with a 3 log reduction but ZOI. The barrier test is a simple method to undertake and can determine if the microorganism translocates from the inner surface (bottom) of the dressing to the outer surface (top), thus demonstrating a barrier effect. This method can also be used to see if the microorganism can translocate from the outer surface to the inner surface of the dressing. The results of most *in-vitro* methods is highly dependent upon the culture medium, the incubation temperature, the amount of inoculum used and moisture present in the test system. When interpreting *in-vitro* method these factors should be taken into account. The antimicrobial effect of silver can be compromised where there are high levels of protein because the salts bind and form inactive complexes. This is also the case if there are large volumes of chloride or sulphate ions.

The antimicrobial and anti-inflammatory effect of silver preparations and dressings has been extensively assessed using murine, guinea pig and porcine models [20, 24, 25] and mostly show a positive effect on wound healing.

*In-vitro* assessment of wound healing using tissue culture of representative cell lines *e.g.* keratinocytes and fibroblasts, frequently demonstrates cytotoxic activity [31]. As these methods use single cells rather than the complex architecture in the epidermis and dermis, these results should be interpreted very carefully [32]. Multi-laminate structures resembling the epidermis can be grown in the laboratory and these can be used to examine the toxicity of silver and the MIC.

Acticoat 7$^{TM}$ and Contreet$^{TM}$ foam are effective wound care products and have been used for a number of years without reported toxic risk [33, 34]. Laboratory animal studies can be used to predict benefits or toxicity of silver-containing products on wound repair and regeneration, but results should be extrapolated carefully, as the local inflammatory and systemic response can vary depending upon the animal used and can only produce preliminary information. Human skin is unique in the animal kingdom and has no exact counterpart, even among primates [35].

## 6. TOXICITY OF SILVER

After 50 years of intensive use in burns and in recent years in chronic wounds, reported cases of silver toxicity are limited. Documented cases of argyria in burns were reported in the 1960s following widespread use of silver nitrate soaked with uncontrolled levels of silver following frequent dressing changes. Subsequently there has only been an occasional report and these have been associated with ingested silver with arygria being the most common manifestation [21, 36, 37]. There have been reports of some patients experiencing stinging and localised pain on application but this resolves relatively quickly. Sensitization can occur albeit rarely and at a lower incidence to other topical antimicrobials used in wound dressings.

Silver levels absorbed across intact skin were barely above normal in healthy volunteers [38] compared to levels absorbed across open wounds and serum Ag levels peaked at 6hrs. Urine levels were shown to be 100/400 times higher than serum levels [39]. The maximum serum Ag level of a cohort of patients (n=30) treated with nanocrystalline silver dressings was 230 micrograms/ml with a median of 56.8, and correlated with the size of the burn and length of exposure to silver. It was also shown that within 6 months post treatment, serum silver levels had returned to normal [40].

*In-vitro* studies using cultured keratinocytes or fibroblasts which are important in wound closure have demonstrated cyto-toxicity [41] but the *in-vitro* findings do have be judged against the effects seen in the patient.

## 7. CASE STUDIES AND HEALTH ECONOMICS

The use of silver dressings has increased rapidly over the last five years and with it huge associated costs. A 20 fold increase in spending on silver dressings in the NHS in the UK was reported in 2007 (£858k in 2004; compared to £23m in 2006) [5]. Recent studies have tried to assess quality of life issues, comfort, and time to healing and economic impact of silver dressings compared to conventional treatment without any antimicrobial agent or an alternative agent, for example, iodine or honey. The results certainly need to be discussed in further detail and their needs to be a more comprehensive assessment of outcomes in order to make an informed decision. Many different parameters need to be included if using a health economic argument in addition to the dressing cost including length of stay, no of dressing changes, pain during dressing change and quality of life in terms of odour control and comfort.

In a recent randomized control trial undertaken in the UK (VULCAN study) over a 30 month period (March 2005-November 2007), comparing a variety (n=6) of silver dressings to compression therapy for treatment of leg ulcers, with an end point of healing (12 weeks), no significant difference was obtained between the two study groups. However, the design and finding of the study is questionable as silver dressings were used for the duration of the study period whether or not they were clinically necessary. This study obviously concluded that the cost of silver dressings was much higher than the control group with the same outcome [42]. However, in a meta-analysis carried out by Lo *et al.* [43] on the use of silver dressings on chronic wounds, their findings statistically confirmed the effectiveness of silver dressings when time to wound healing and improvement in patients' quality of life was included in the meta-analysis.

Cost effectiveness is an interesting problem to be addressed when choosing any wound dressing. The cost associated with the use of nanocrystalline silver dressings has been studied on a number of occasions in a number of countries worldwide. Studies evaluating the cost effectiveness, pain, time to healing and rate of infection in burns have shown that the newer nanocrystalline silver dressings are more effective than previous forms of silver in creams and ointments [44]. Costs can be therefore be justified if there is definite benefit to the patient in terms of reduced dressing changes, improved time to healing, reduced infection rates and associated surgical interventions and most importantly improved quality of life for the patient. A systematic audit on the use of nanocrystalline $Ag^+$ dressings as a preventative treatment on burns compared conventional treatment with a SSD / chlorhexidine cream showed a reduction in cost because reduced dressing changes, nursing time, prevention of infection and the use of antibiotics [45].

It has been suggested that prolonged usage of the antimicrobial silver dressings may create more problems in terms of wound healing. However, at this time, there is more evidence to suggest the benefits of nanocrystalline silver dressings in acute wounds (burns, toxin mediated and immunological) and chronic wounds (pressure ulcers, venous and diabetic foot ulcers), [46, 47] rather than disadvantage. Concerns have been raised to developing resistance to silver is through prevention of entry into the cell or through efflux mechanisms but the reported cases of genuine silver resistance is minimal and provided appropriate levels are used as seen with nanocrystalline silver, the numbers should be kept at a minimum [48].

# REFERENCES

[1]     Moyer CA. A treatment of burns. Trans Stud Coll Physicians 1965; 33: 53-103.

[2]     Moyer Ca, Brentano L, Gravens Dl *et al.* Treatment of large human burns with 0.5% silver nitrate solution. Arch Surg 1965; 90: 812-67

[3]     Fox CL. Silver sulfadiazine-a new topical therapy for Pseudomonas in burns. Arch Surg 1968; 96: 184-8.

[4]     Cutting K, White R, Hoekstra H. Topical silver impregnated dressings and the importance of dressing technology Int Wound J 2009; 6: 396-402

[5]     Chambers H, Dumville JC, Cullum N. Silver treatments for leg ulcers: a systematic review. Wound Repair Regen 2007; 15: 165-73.

[6]     Richards RME, Taylor RB, Xing DKL. An evaluation of antibacterial activities of sulphonamides, trimethoprim, dibromopropamidine and silver nitrate compared with their uptakes in selected bacteria J Pharmaceutical Science 1991; 80: 861-867.

[7]     Dollery C. Silver sulphadiazine; 1991, In Therapeutic Drugs, Churchill Livingstone.

[8]     Dunn K, Edwards-Jones V. The role of Acticoat TM with nanocrystalline silver in the management of burns. Burns 2004; 30 (Suppl 1): S1–9.

[9]     Pal S, Tak YK, Song JM. Does the antibacterial activity of the silver nanoparticles depend on the shape of the nanoparticles? A study of the gram-negative bacterium Escherichia coli. Appl Environ Microbiol 2007; 73: 1712-1720.

[10]    Zhao Y, Jiang Y Fang Y. Spectroscopy of silver nanoparticles Spectrochim Acta 2006; 65: 1003-1006

[11]    Hamilton-Miller JM, Shah S, Smith C. Silver sulphadiazine; a comprehensive in-vitro reassessment Chemotherapy 1993; 39: 405-409.

[12]    Schreurs WJ and Roseberg H. Effect of silver ions on transport and retention of phosphate by Escherichia coli. J Bacteriol 1982; 152: 7-13.

[13]    Dibrov P, Dzioba J, Gosink KK *et al.* Chemiosmotic mechanism of antimicrobial activity of $Ag^+$ in Vibrio cholerae Antimicrobial Agents Chemotherapy 2002; 46: 2668-2670.

[14]    Chappell JB, Grenville GD. Effect of silver ions on mitochondrial adenosine triphosphatase. Nature 1954; 174: 930-1.

[15]    Semeykina AL, Skulachev VP. Submicromolar $Ag^+$ increases passive $Na^+$ permeability and inhibits respiration-supported formation of $Na^+$ gradient in Bacillus FTU vesicles. FEBS Lett 1990; , 269: 69-72.

[16]    Modak SM, Fox CL Jr. Binding of silver sulfadiazine to the cellular components of Pseudomonas aeruginosa. Biochem Pharmacol 1973; 22: 2391-404

[17]    Rosenkranz HS and Rosenkranz S. Silver sulfadiazine: interaction with isolated deoxyribonucleic acid. Antimicrob Agents Chemother 1972; 2: 373-83

[18]    Teng QL, Wu J, Chan GQ *et al.* Mechanistic study of the antibacterial effect of silver ions on Escherichia coli and Staphylococcua aureus. J Biomed Mater Res 2000; 52: 662-668.

[19]    Lansdown ABG. Silver: its antibacterial properties and mechanism of action. J Wound Care 2002; 11: 173.

[20]    Wright JB, Lam K, Buret AG *et al.* Early healing events in a porcine model of contaminated wounds: effects of nanocrystalline silver on matrix metalloproteinases, cell apoptosis, and healing. Wound Repair Regen 2002; 10: 141-51.

[21] Lansdown AB. Silver in health care: antimicrobial effects and safety in use. Curr Probl Dermatol 2006; 33: 17-34.

[22] Asz J, Asz D, Moushey R *et al.* Treatment of toxic epidermal necrolysis in a pediatric patient with a nanocrystalline silver dressing. J Pediatr Surg 2006; 41: 9-12.

[23] Dalli RL, Kumar R, Kennedy P *et al.* Toxic epidermal necrolysis/Stevens-Johnson syndrome: Current trends in management. ANZ J Surg 2007; , 77: 671-6.

[24] Bhol KC, Alroy J, Schechter PJ. Anti-inflammatory effect of topical nanocrystalline silver cream on allergic contact dermatitis in a guinea pig model. Clin Exp Dermatol 2004; 3: 282-7.

[25] Bhol KC, Schechter PT. Topical Nanocrystalline cream suppresses proinflammatory cytokines and induces apoptosis of inflammatory cells in murine model of allergic contact dermatitis Br J Dermatol 2005; 152: 1235-1242

[26] Nadworthy PL, Wang J, Tredget EE *et al.* Anti-inflammatory activity of nanocrystalline silver in a porcine contact dermatitis model. Nanomedicine 2008; 4: 241-251

[27] Thomas S, McCubbin P. A comparison of the antimicrobial effects of four silver-containing dressings on three organisms. J of Wound Care 2003; 12: 3; 101-107)

[28] Gallant-Behm CL, Yin HQ, Liu S *et al.* Comparison of in-vitro disc diffusion and time kill assays for the evaluation of antimicrobial wound dressing efficiency. Wound Repair Regeneration 2005; 13: 412-417

[29] Strohal R, Schelling M, Takacs M *et al.* Nanocrystalline silver dressings as an efficient anti-MRSA barrier: a new solution to an increasing problem. J Hosp Infect 2005; 60: 226-30

[30] Edwards-Jones V. Antimicrobial and barrier effects of silver against methicillin-resistant Staphylococcus aureus. J Wound Care 2006; 15: 285-90.

[31] Hollinger MA. Toxicological aspects of topical silver pharmaceuticals. Crit Rev Toxicol 1996; 26: 255–60.

[32] Lansdown AB, Williams A. How safe is silver in wound care? J Woundcare 2004; 13: 131-136.

[33] Demling RH, DeSanti L. The rate of re-epithelialization across meshed skin grafts is increased with exposure to silver. Burns 2002; 28: 264

[34] Karlsmark T, Agerslev RH, Bendz SH *et al.* Clinical performance of a new silver dressing, Contreet Foam, for chronic exuding venous leg ulcers. J Wound Care 2003; 12: 351-4.

[35] Lansdown ABG. Animal models for the study of skin irritants. Curr Probl Dermatol 1978; 7: 26–38.

[36] Payne CM, Bladin C, Colchester AC *et al.* Argyria from excessive use of topical silver sulphadiazine. Lancet 1992; 340: 126

[37] Tomi NS, Kränke B, Aberer W. A silver man. Lancet 2004; 363: 532

[38] Wan AT, Conifers RT, Coombes CJ *et al.* Determination of silver in blood, urine and tissues of volunteers and burn patients. Clin Chem 1991; 37: 1683-1687.

[39] Coombes CJ, Wan AT, MAsterton JP *et al.* Do burns patients have a silver lining? Burns 1992; 18: 179-184.

[40] Vlachou E, Chipp E, Shale E *et al.* The safety of nanocrystalline silver dressings in burns. A study of systemic silver absorption. Burns 2007; 33: 979-985.

[41] Fraser JF, Cuttle L, Kempf M *et al.* Cytotoxicity of topical antimicrobial agents used in burn wounds in Australasia. ANZ J Surg 2004; 74: 139–212.

[42] Michaels JA, Campbell B, King B *et al.* Randomized controlled trial and cost-effectiveness analysis of silver-donating antimicrobial dressings for venous leg ulcers (VULCAN trial). Br J Surg. 2009; 96: 1147-56.

[43] Lo SF, Chang C-J, Hu W-Y *et al.* The effectiveness of silver-releasing dressings in the management of non-healing chronic wounds: a meta-analysis. J Clin Nursing 2009; 18: 716–728

[44] Paddock HN, Fabia R, Giles S *et al.* A silver-impregnated antimicrobial dressing reduces hospital costs for pediatric burn patients J Pediatric Surg 2007; 42: 211–213

[45] Fong J, Wood F, Fowler B. A silver coated dressing reduces the incidence of early burn wound cellulitis and associated costs of inpatient treatment: comparative patient care audits. Burns 2005; 31: 562-7.

[46] Wright JB, Lam K, Burrell RE. Wound management in an era of increasing bacterial antibiotic resistance: a role for topical silver treatment. Am J Inf Control 1998; 26: 572–7.

[47] Ziegler K, Gorl R, Effing J *et al.* Reduced cellular toxicity of a new silver-containing antimicrobial dressing and clinical performance in nonhealing wounds. Skin Pharmacol Physiol 2006; 19: 140–6.[48]

[48] Silver S, Pheung LT and Silver G. Silver as biocides in burn and wound dressings and bacterial resistance to silver compounds. J Indust Microbiol Biotech 2006; 33: 627-634.

# CHAPTER 4

# Nanomedicine and the Treatment of Coronary In-Stent Restenosis – A Clinical Review

## Garry McDowell[*]

*Faculty of Health and Social Care, Edge Hill University, UK*

**Abstract:** Coronary in-stent restenosis remains a significant limitation to the long term efficacy of coronary artery stent placement. In this Chapter the author reviews the pathophysiology of coronary in-stent restenosis, together with an overview of the current treatment modalities. The potential for the use of nanotechnology is also reviewed.

The first human safety trial of systemic nanoparticle paclitaxel (nab-paclitaxel) for in-stent restenosis (SNAPIST-I) is discussed. The results showed no significant adverse advents attributable to the nab-paclitaxel at 10 or 30 $mg/m^2$, although moderate neutropenia, sensory neuropathy and mild to moderate reversible alopecia occurred at higher doses. No major adverse cardiac events were recorded at 2 months, whilst at 6 months 4 target lesions required revascularisation. The investigators concluded therefore that systemic nab-paclitaxel was well tolerated at a dose of less that 70 $mg/m^2$. To date however no formal clinical evaluation has been reported as to the clinical utility of nab-paclitaxel or any of the nano preparations discussed for the suppression of coronary in-stent restenosis.

**Keywords:** Nanoparticles; coronary in-stent restenosis; pathophysiology; treatment; animal models; human; clinical trials; vascular disease; atherosclerosis; clinical review.

## 1. INTRODUCTION

Nanotechnology is a relatively new science concerning the use of nanoscale structures (0.1-100 nm) and is derived from the Greek word meaning dwarf. One nanometre ($10^{-9}$) equals one billionth of a metre. Nanomedicine is the medical applications of nanotechnology to the study of human disease [1]. In this chapter we will examine the current role of nanomedicine, from a clinical perspective, in the treatment of coronary in-stent restenosis.

## 2. INTRODUCTION TO NANOPARTICLES

The major limitation of many therapeutic agents is targeting the agent to the required site of action, for example the vascular endothelium or a tumour. Nanoparticles may present a novel and innovative approach for targeted drug delivery as they can be targeted to the site of action either actively or passively. Active targeting can be achieved by the use of a cell specific ligand, while passive targeting may be achieved by the use of macromolecules such as high molecular weight polymers that target highly vascular areas such as tumours [2,3].

Alternatively nanoparticle suspensions may be infused directly into the required area, providing sufficient access (eg venous or arterial) can be gained [4]. A number of biocompatible nanomaterials are available which have favourable biological, chemical and physical characteristics [1], for example:

- Long circulating half life

- Selective binding to cells and tissues

- Sufficient signal-to-noise ratio for imaging

*Address correspondence to Garry McDowell: Senior Lecturer, Faculty of Health, Edge Hill University, St Helens Road, Ormskirk, Lancashire, L39 4QP, UK; Tel: 01695 650720; E-mail: garry.mcdowell@btinternet.com

**Mark Slevin (Ed)**

- Low toxicity

- Easy to produce

- Easy to detect with current imaging modalities

Nanoparticles are ideal drug delivery systems. Their small size means they can penetrate through the capillary network. In addition, they are readily taken up by cells permitting efficient drug accumulation in the target tissues. Moreover nanoparticle composition allows for sustained release of the pharmacological agent over a period of time [5].

## 3. TYPES OF NANOPARTICLES

### 3.1 Polymer Biodegradable Nanoparticles

These are solid, colloidal particles of macromolecules that range in size from 10-1000nm [1]. They are idea drug delivery systems [6], where the compound of interest may be dissolved, entrapped, adsorbed, attached or encapsulated into the nanoparticle matrix [7,8]. There are two classes in this category [1]:

- Nanocapsules are vesicular in which the drug is confined to a cavity surrounded by a polymer membrane

- Nanospheres the drug is physically and uniformly distributed throughout the nanosphere matrix.

### 3.2 Ceramic Nanoparticles

These have a number of advantages in that they can be prepared relatively easily with desired shape, size and porosity. Their small size (<50nm) is highly advantageous permitting them to escape the reticuloendothelial system. Ceramic nanoparticles provide good protection to molecules enclosed within them. In addition their surface can be modified to facilitate targeted drug delivery [9,10].

### 3.3 Micelles

Micelles are characterised by their core-shell structure, with hydrophobic segments surrounded by a hydrophilic shell, and are particularly useful for the delivery of hydrophobic drugs [11]. Micelles have a small size, typically in the range <100 nm which makes them ideal drug delivery systems since they can evade the reticuloendothelail system [6]. Site directed targeting may be facilitated by conjugating ligands to the micelle surface [12].

### 3.4 Liposomes

Liposomes are small and have a spherical shape and are formed from natural and non toxic phospholipids and cholesterol. As liposomes are small and posses hydrophobic and hydrophilic characteristics there are ideally suited to the development of novel drug delivery systems [13]. Liposome surfaces can be modified to increase circulating half life and conjugated to antibodies or ligands for enhanced tissue specificity.

### 3.5 Dendrimers

These are macromolecules that comprise a series of branches around an inner core [14]. As such they can display multiple surface groups for biomedical applications. Biological interaction with dendrimers is *via* terminal groups which are capable of modification for site specific targeting [15].

### 3.6 Quantum Dots

These are composed of a crystalline structure of a few hundred atoms that form a core. The core is then covered by an outer insulating material. This unique structure means that when a photon of visible light hits the crystal core, the energy is confined to the core before being released as fluorescence [16].

### 3.7 Magnetic Nanoparticles

Magnetic nanoparticles can be used as contrast agents in Magnetic Resonance Imaging (MRI Scans). Magnetic nanoparticles are composed of an iron oxide core covered with dextran or silicon [17].

## 4. OVERVIEW OF NANOMEDICINE APPLICATIONS IN CARDIOLOGY

Cardiovascular disease, including acute coronary syndromes and cerebrovascular events continue to be a major source of mortality and morbidity. Current medical screening and diagnosis is limited and many of the symptoms and signs of cardiovascular and cerebrovascular disease are non-specific. Data from our own group examining patients presenting to the emergency department has shown several 'atypical' symptoms actually render AMI more likely, whereas many 'typical' symptoms that are often considered to identify high-risk populations have no diagnostic value [18].

Nanomedicine provides a unique opportunity to explore at a cellular or organ level the various pathophysiologies of the cardiovascular system. Nanomolecules have been used in [19]:

- Assessing and treating atherosclerosis in asymptomatic patients

- Coronary revasculariation

- Thrombolytic therapy

- Treatment of coronary in-stent restenosis.

## 5. CORONARY IN-STENT RESTENOSIS

Coronary in-stent restenosis has been reviewed at length in a number of detailed reviews [20-22]. Percutaneous Transluminal Coronary Angioplasty (PTCA) is a minimally invasive technique for the treatment of coronary artery atherosclerosis. The benefits of PTCA however are limited by restenosis occurring in 5-35% of patients within the first year of treatment [23].

### 5.1 Pathophysiology of Coronary In-Stent Restenosis

The pathophysiology of in-stent restenosis is multi-factoral and is a result of vascular injury caused by shear stress associated with balloon inflation and stent placement. It can briefly be summarised below (Fig. **1**):

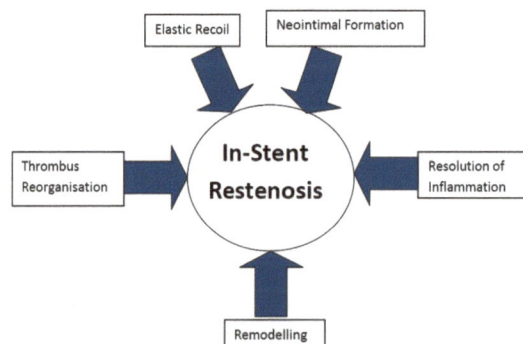

### *5.1.1 Elastic Recoil*

Stent implantation results in stretching of the elastic fibres that make up the internal and external elastic laminae. Elastic recoil of the overstretch fibres follows after balloon deflation and can result in up to 40% loss of luminal area, although stents reduce elastic recoil [22,23].

### 5.1.2 Thrombus Reorganisation

PTCA results in endothelial denudation and exposure of procoagulant compounds such as collagen, von Williebrand factor and fibronectin. Previous studies have shown that fibrin and platelets are deposited at restenotic sites suggesting that mural thrombus formation promotes in-stent restenosis [22,23].

### 5.1.3 Neointimal Formation

PTCA induced arterial injury promotes Vascular Smooth Muscle Cell (VSMC) proliferation and migration as a result of:

- Mechanical stretch

- Endothelial denudation

- Cytokine release from endothelial cells, VSMCs and inflammatory cells.

Neointimal formation is a major cause of in-stent restenosis and increases up to 3 months after PTCA, with little change at 6 months with gradual reduction between five months and 3 years.

Neointimal formation after stenting is associated with medial disruption proliferation and does not correlate with time of injury. Indeed most of the proliferating cells are located next to the stent struts, leading authors to suggest that proliferation is a chronic low grade reaction to the stent.

Interestingly restenotic lesions are hypocellular compared to primary atherosclerotic plaques, instead consisting of mainly collagen and matrix proteoglycan. The cellular component comprises only 10% of the tissue mass in post PTCA neointima [22,23].

In addition, neointimal formation within a stent results from axial movement of primary plaque displaced to adjacent artery segments by the PTCA procedure [22,23].

### 5.1.4 Remodelling

Negative remodelling is a major cause of coronary PTCA restenosis, although its mechanism remains unknown. Studies using intravascular ultrasound have shown that remodelling causes between two thirds and three quarters of the lumen loss associated with in-stent restenosis. Remodelling is largely mitigated by stent placement [22,23].

### 5.1.5 Resolution of Inflammation

Vascular inflammation has been identified as an important in the pathophysiology of coronary in-stent restenosis and several inflammatory biomarkers have been identified as potential predictors of the development of in-stent restenosis [21].

## 5.2 Treatment and Prevention of In-Stent Restenosis

As the pathophysiology of in-stent restenosis is multifactorial a number of therapeutic strategies are currently employed and these will be described briefly below. The two main classes are:

- Brachytherapy

- Drug eluting stents

### 5.2.1 Brachytherapy

Intra-coronary brachytherapy can be delivered using either β or γ sources, and has been used as a successful therapeutic strategy, although a recent long term follow up study, conducted over a period of approximately

39 months has shown delayed, though continued, restenotic process after index procedure. Brachytherapy reduces neointimal accumulation, by blocking cell proliferation, inducing apoptosis and inhibiting cell migration [24,26].

Brachytherapy however has significant limitations including late thrombosis resulting in myocardial infarction and paradoxical restenosis at the ends of the stent [27]. Further recent investigation has shown that sirolimus eluting stent remains superior to intra-coronary β-radiation for treating diffuse in-stent restenosis [28].

### 5.2.2 Drug Eluting Stents

Since the first stent was implanted in 1986, stent technology has rapidly progressed, with the first drug eluting stents being commercially available in 2002, with 2 main classes of drug have been used [29].

### 5.2.2.1 Rapamycin (Sirolimus)

Sirolimus possess weak antibiotic activity, together with immune modulating activity directed against both T-lympocyte activation and cell proliferation. From the above discussion of the pathophysiology of in-stent restenosis, sirolimus may act by a number of different mechanisms to limit coronary in-stent restenosis [29]. Sirolimus:

- Is a potent inhibitor of human Vascular Smooth Muscle Cell (VSMC) proliferation

- Reduces VSMC migration.

- Reduces cell matrix growth. As intima forms a major part of the restenotic lesion, this affect has been suggested to be more important than the antiproliferative effect.

- Is an inhibitor of inflammation in mammalian blood vessels after injury.

The above biological actions mean that rapamycin drug eluting stents are potent inhibitors of neointimal formation in both animal models and clinical studies and low levels of restenosis for up to 2-3 years [30].

### 5.2.2.2 Paclitaxel

The taxane class of drugs are potent antiproliferative agents used in the treatment of cancer. Paclitaxel promotes the polymerisation of the α and β subunits of tubulin, by reversibly binding to the β subunit, causing stabilisation of the microtubules and inhibition of cell division and proliferation. In addition, tubulin is also important in various cellular functions including axonal transport. Administration of paclitaxel causes total growth inhibition within the dose range 1-10 mmol/L, but is toxic at higher concentrations [29].

The two most successful commercially available drug eluting stents are the sirolimus eluting Cypher stent[©] (Cordis, Johnson and Johnson, Miami, FL) and the paclitaxel eluting Taxus stent™ (Boston Scientific, Natick, MA). Clinical trail data support the use of these stents for the reduction of coronary in-stent restenosis [30-33].

## 6. THE USE OF NANOTECHNOLOGY FOR THE TREATMENT OF CORONARY IN-STENT RESTENOSIS

There is still a significant requirement for a novel drug delivery mechanism for the treatment of coronary in-stent restenosis, due to the limitations of the current modalities including late stent thrombosis. A nanoparticle based approach is ideal for the treatment of restenosis since targeted delivery of nanoparticles is feasible and much lower concentrations of the active drug can be used hence reducing systemic toxicity.

The size of particle however is critical in the distribution of nanoparticles in the blood vessel wall. Westedt *et al.* [34], in experiments conducted using the aorta abdominalis of New Zealand White Rabbits as a model system, report that nanoparticles of 100 and 200 nm are able to penetrate to the inner layers of the vessel wall, while 514nm particles accumulate predominantly at the luminal surface. Some examples of nanoparticles used in the treatment of in-stent restenosis are reviewed below.

## 6.1 Lipid Based Nanoparticles

Lipid based nanoparticles have been utilised to deliver a number of different classes of drug to the arterial endothelium.

Clodronate, a bisphonsphonate, has been delivered using liposome nanoparticle of 1:3 distearoyl phosphatidylglycerol, 1, 2 distearoly –sn-gylcero-3-phosphcholine [35]. Lyposomal choldronate inhibited neointimal growth in the balloon injured rabbit carotid artery after systemic administration. Other members of the bisphosphonate class of drugs including pamidronate and alendronate have been utilised as antirestenotic agents in balloon injured rat carotid artery model [36]. It is noteworthy, however that these experiments were performed in a carotid artery model and whether these results are relevant to coronary restenosis after PTCA remains unknown.

TRM 484 consists of nanoparticles of prednisolone with high infinity to chrondroitin suphate proteoglycans and at a dose of 1mg/kg significantly reduced neo intimal growth in atherosclaratic New Zealand white rabbits implanted with bare metal stents [37].

## 6.2 Polymeric Based Nanoparticles

Early work on polymeric nanoparticles began with a comparison of probucol delivery by polymeric and liposomal nanoparticles. Probucol has been shown to reduce restenosis after angioplasty provided oral administration is commenced one month before the procedure [38]. Klughertz and colleagues prepared $^{35}$S-probucol incapsulated in either liposomal or Polylactic –Coglycolic Acid (PLGA) nanoparticles, which were delivered using and infusion catheter after balloon angioplasty of rabbit iliac arteries. Iliac arteries, perivascular fat and downstream tissues were harvested and the radioactivity measured from animals euthanized on day 0,3 and 7 after dosing. The results showed after delivery efficiency was superior with PLGA [38]. It should be noted however that these experiments were conducted outside the coronary circulation.

Cohen Sela *et al.* reported that PLGA nanoparticles containing alendronate reduced neeintimal formation and restenosis by systemic transient depletion of monocytes in hypercholesterolaemic rabbit model [39]. Further work by the same group reported incorporation of the bisphosphonates, 2- (2-Aminopyrimidino) ethyldiene -1, 1-bisphosphonbic acid betaine (ISA) into PLGA based nanoparticles (ISA – NP) [40]. Intravenously administered ISA-NP resulted in a significant attenuation of restenosis by 45%, 14-days after carotid artery injury in comparison to a control group of animals treated with free ISA, buffer or blank nanoparticles. However the effect was not preserved long-term (30-days post injury) and no significant reduction in neointimal reduction was observed. Surprisingly, significant neointimal suppression was observed following subcutaneous injection of ISA-NP [40].

Paclitaxel is a member of the taxane family of drugs. Paclitaxel loaded nanoparticles have been prepared from oil-water emulsion using biodegradable PLGA and surface modified with the cationic surfactant Didoceylmethylammonium Bromide (DMAB) to enhance arterial retention. *In vivo* investigations have been performed in balloon injured rabbit carotid arteries treated with a single infusion of paclitaxel loaded nanoparticles and observed for 28 days. The results demonstrated that the inhibitory effect on intimal proliferation was dose dependant, and at 30 mg/ml nanoparticle concentration, completely inhibited intimal proliferation, leading the group to speculate that the surface modified paclitaxel loaded nanoparticles provide an effective means of inhibiting the proliferative response to vascular injury [41]. Further results by Westedt *et al.* [42] substantiated these findings utilising paclitaxel loaded nanoparticles administered locally to the wall of balloon injured rabbit iliac arteries using a porous balloon catheter. The results demonstrated a 50% reduction in neointimal area compared to the control vessels treated with blank nanoparticles.

Work from another group has demonstrated that the antiproliferative effects of Paclitaxel can be significantly improved by co-administration of other agents [43]. C6-ceramide is an apoptotic signally molecule and has been combined with Paclitaxel in polymeric nanoparticles consisting of poly( ethylene oxide) – modified poly (episilon caprolactone). Combination of Paclitaxel with cereamide when administered in nanoparticle formulation significantly augmented the antiproliferative effect of either agent alone [43].

The angiotensin-converting-enzyme inhibitor, lisinopril , has also been encapsulated in nanoparticles of PLGA for site specific delivery by catheters for the prevention of coronary in-stent restenosis [44], although to date *in vivo* studies to examine the anti restenotic effect have not been reported.

Further work from Cohen-Sela's group [45] incorporated the antiproliferative agent mitramycin into PLGA nanoparticles using a nanoparticiption technique. Unfortunately *in vivo* testing using a rat carotid artery model showed no inhibition of restenosis. The authors suggest that this is probably due to the short depletion period of circulating monocytes and the lack of arterial targeting.

Work to increase the bioadhesive properties of nanoparticles has been suggested to improve retention and arterial uptake of nanoparticles into the arterial wall [46]. Zou *et al.* [46] prepared bioadhesive PLGA nanoparticles, encapsulating rapamycin, using different concentration of carbopol 940, however *in-vivo* results are awaited.

Recent research has focused on the administration of drugs using biodegradable polymer nanoparticles capable of prolonged drug release. Sustained drug release of dexamethasone or rapamycin from nanoparticles based on poly (ethylene oxide) and poly (d.L-lactic-co-glycolic acid) block copolymers has been investigated [47]. The investigators found that treating the nanoparticles with gelatine or albumin after drug loading resulted in a linear drug release, the rate of release being related to the amount of protein associated with the nanoparticles [47]. Release of dexamethasone and rapamycin was sustained for 17 and 50 days respectively [47].

Luderer *et al.* [48] report the use of sirolimus loaded biodegradable poly (D,L lactide) nanoparticles as drug carriers to prevent restenosis following coronary angioplasty [48]. The particles showed a biphasic release pattern consisting of a short burst release of 50% w/w sirolimus followed by a longer slower release.

Moreover, Nakano *et al.* [49] have succeeded in formulating a nanoparticle eluting stent. In a porcine coronary artery model, the magnitude of stent induced injury, inflammation, endothelial recovery and neointimal formation were comparable between bare metal stent and nanoparticle eluting stent. It is worthy of note however that the study does not present any data on restenosis rate, or the incorporation of any pharmaceutical preparations within the nanoparticles.

The tryphostins are a class of Platelet Derived Growth Factor (PDGF) receptor $\beta$ tyrosine kinase blockers [50,51]. Preclinical investigations have reported results with the experimental compound AG-1295 incorporated in polylactide nanoparticles. PLA AG-1295 nanoparticles were delivered *via* an infusion catheter in a balloon injured swine model, resulting in inhibition of smooth muscle cell (SMC) growth. Further, another tyrphostin AGL-2043 encapsulated in in PLA nanoparticles inhibited restenosis in both balloon injured rat carotid artery and stented porcine artery models [52].

## 6.3 Gel Like Nanoparticles

Previous research [53] has demonstrated that nanosized (100 nm) hydrogel spheres made of poly(N-isopropylacrylamide) are internalised by endothelial cells and VSMC more than microspheres (1μm), although cellular uptake was dependant on the incubation time, sphere concentration and introduced shear stress levels of the medium. In contrast microspheres were rapidly taken up be phagocytes, especially at high concentration [53]. These findings lead the authors to suggest that hydrogel nanospheres are more effective as an intravascular delivery system in terms of vascular uptake and biocompatibility [53].

Since significant number of VSMC undergo rapid apoptosis following balloon angioplasty Reddy and colleagues [54] tested the hypothesis that preventing VSMC from apoptosis could prevent intimal hyperplasia. They used rapamycin (which has anti-apoptotic and antiproliferative actions) loaded gel nanoparticles of mean diameter 54 nm. When infused into a rat carotid artery model of vascular injury the authors report significant inhibition of hyperplasia and re-endothelialisation of the injured artery. Further, the group report inhibition of activation of caspase 3/7 enzyme systems in the treated artery, preventing VSMC from undergoing apoptosis [54].

### 6.4 Miscellaneous Studies

Kolodgie *et al.* [55] report the preparation of Paclitaxel loaded albumin based nanoparticles for the reduction of in-stent neointimal growth. The research conducted in New Zealand White rabbits receiving bilateral iliac artery stents yielded significant results. Systemic administration of albumin nanoparticles containing Paclitaxel reduced neointimal growth at 28 days. A further single repeated dose was required or sustained neointimal suppression at 90 days post procedure [55]. Moreover further pre-clinical work has demonstrated the utility of tissue factor targeted nanoparticles containing doxorubicin or Paclitaxel to inhibit VSMC proliferation in culture [56].

In addition, intra mural delivery of αVβ3-targeted rapamycin loaded nanoparticles inhibited stenosis without delaying endothelial healing after balloon injury [57].

Chorney *et al.* [58] report the use of uniform magnetic fields to control the release of Paclitaxel from biocompatible magnetic nanoparticles. The research showed that magnetic treatment of cultured arterial SMC with Paclitaxel loaded magnetic nanoparticles caused significant inhibition of cell growth, which was not observed under non-magnetic conditions. The authors postulate that the results demonstrate the feasibility of site specific drug delivery by uniform field controlled targeting of magnetic nanopatticles [58].

### CONCLUSIONS

Much work therefore has been undertaken to evaluate the potential clinical utility of nanoparticles for the targeted and non-targeted delivery of various agents with antiproliferative and anti-restenostic actions. To date most of these investigations have been conducted either *in vitro* or *in vivo* utilising animal models. Many studies have also been conducted outside the coronary circulation and hence the relevance of the result to coronary in-stent restenosis can only be postulated.

In 2007 however, Margolis *et al.* [59] presented the first human safety trial of systemic nanoparticle Paclitaxel (Nab-Paclitaxel) for in-stent restenosis I (SNAPIST-I). In this study the investigators administered systemic treatment with a 130nm albumin nanoparticle encapsulating paclitaxel in 10, 30, 70 and 100 $ng/m^2$ intravenously after stenting of a single lesions of $\geq$ 3 mm in 23 patients. The results showed no significant adverse advents attributable to the nab-paclitaxel at 10 or 30 $mg/m^2$, although moderate neutropenia, sensory neuropathy and mild to moderate reversible alopecia occurred at higher doses. No major adverse cardiac events were recorded at 2 months, whilst at 6 months 4 target lesions required revascularisation. The investigators concluded therefore that systemic nab-paclitaxel was well tolerated at a dose of less that 70 $mg/m^2$ [23]. To date however no formal clinical evaluation has been reported as to the clinical utility of Nab-Paclitaxel for the suppression of coronary in-stent restenosis.

### REFERENCES

[1]     Patel D, Bailey SR. Nanotechnology in cardiovascular medicine. Cath Cardiovasc Interv 2007;069: 643-54.
[2]     Maeda H. The enhanced permeability and retention effect in tumour vasculature. The key role of tumour selective macromolecular drug targeting. Adv Enzyme Reg 2001; 41: 189-207.
[3]     Panyam J, Lof J, O'Leary E *et al.* Efficiency of dispatch and infiltrator cardiac infusion catheters in arterial localisation of nanoparticles in a porcine coronary model of restenosis. J Drug Target 2002; 10: 515-23.
[4]     Guzman LA, Labhasetwar V, Song C *et al.* Local intramural infusion of biodegradable polymeric nanoparticles. A novel approach for prolonged drug delivery after balloon angioplasty. Circulation 1996; 94: 1441-8.

[5]     Lanza GM, Yu X, Winter PM *et al.* Targeted antiproliferative drug delivery to vascular smooth muscle cells in a magnetic resonance imaging contrast agent: Implications for rational therapy of restenosis. Circulation 2002; 106: 2842-7.

[6]     Nakanishi T, Fukushima S, Okamoto K *et al.* Development of the polymer micelle carrier system for doxorubicin. J Control Release 2001; 74: 295-302.

[7]     Torchilin VP. Targeted polymeric micelles for delivery of poorly soluble drugs. Cell Mol Life Sci 2004; 61: 2549-59.

[8]     Torchilin VP. Structure and design of polymeric surfactant based drug delivery systems. J Control Release 2001; 73: 137-72.

[9]     Roy I, Ohulchansky TY, Pudavar HE *et al.* Ceramic based nanoparticles entrapping water insoluble photosensitising anticancer drugs: A novel drug carrier system for photodynamic therapy J Am Chem Soc 2003; 125: 7860-5.

[10]    Cherian AK, Rana AC, Jain SK. Self assembled carbohydrate stabilised ceramic nanoparticals for the parental delivery of insulin. Drug Dev Ind Pharm 2000; 26: 459-63.

[11]    Jones M, Leroux J. Polymeric micelles – a new generation of colloidal drug carriers. Eur J Pharm Biopharm 1999; 48: 101-111.

[12]    Torchilin VP. Targeted polymeric micelles for delivery of poorly soluble drugs. Cell Mol Life Sci 2004; 61: 2549-59.

[13]    Torchilin VP. Recent advances with liposomes as pharmaceutical carriers. Nat Rev Drug Discov 2005; 4: 145-160.

[14]    Svenson S, Tomalia DA. Dendrimers in biomedical applications – reflections on the field. Adv Drug Deliv Rev 2005; 57: 2106-29.

[15]    Yang H, Kao WJ. Dendrimers for pharmaceutical and biomedical applicatiosn. J Biomater Sci Polym Ed 2006; 17: 3-19.

[16]    Yu WW, Chang E, Drezek R *et al.* Water soluble quantum dots for biomedical applications. Biochem Biophys Res Comm 2006; 348: 781-6.

[17]    Gupta AK, Gupta M. Synthesis and surface engineering of iron oxide nanoparticles for biomedical applications. Biomaterials 2005; 26: 3995-4021.

[18]    Body R, Carley S, Wibberley C *et al.* The value of symptoms and signs in the emergent diagnosis of acute coronary syndromes. Rususcitation 2010; 81: 281-6.

[19]    Lanza G, Winter P, Cyrus T *et al.* Nanomedicine opportunities in cardiology. Ann NY Acad Sci 2006; 1080: 451-65.

[20]    Bennett M. In-stent restenosis: Pathology and implications for the development of drug eluting stents. Heart 2003; 89: 218-24.

[21]    Schillinger M, Minar E. Restenosis after percutaneous angioplasty: the role of vascular inflammation. Vasc Health Risk Management 2005; 1: 73-8.

[22]    Kibos A, Campeanu A, Tintoiu I *et al.* Pathophysiology of coronary artery in-stent restenosis. Acute Cardiac Care 2007; 9: 111-9.

[23]    Stone GW, Ellis SG, Cannon L *et al.* Comparison of a polymer based paclitaxel eluting stent with a bare metal stent in patients with complex coronary artery disease: A randomised controlled trial. JAMA 2005; 294: 1215-23.

[24]    Silber S, Popma JJ, Suntharalingam M *et al.* Two year follow up of $^{90}$Sr/$^{90}$Y β-radiation versus placebo control for the treatment of in-stent restenosis. Am Heart J 2005; 149: 689-94.

[25]    Vlachojannis GJ, Fichtlscherer S, Spyridopoulos I *et al.* Intracoronary β-radiation therapy for in-stent restenosis: Long term success rate ad prediction of failure. J Interven Cardiol 2010; 23: 60-5.

[26]    Baierl V, Baumgartner S, Pollinger B *et al.* Three year follow up after strontium-90/yttrium-90 β-radiation for the treatment of in-stent coronary restenosis. Am J Cardiol 2005; 96: 1399-1403.

[27]    Nikas DN, Kalef-Ezra J, Katsouras CS *et al.* Long term clinical outcome of patients treated with β-brachytherapy in routine clinical practice. Int J Cardiol 2007; 115: 183-9.

[28]    Park SW, Lee SW, Koo BK *et al.* Treatment of diffuse In-stent restenosis with drug eluting stents vs intracoronary Beta-raDiation therapy: INDEED Study. Int J Cardiol 2008; 131: 70-7.

[29]    Smith EJ, Jain AK, Rothman MT. New developments in coronary stent technology. J Interven Cardiol 2006; 19: 493-9.

[30]    Kastrali A, Mehilli J, Van Beckerath N *et al.* Sirolimus eluting stents or paclitaxel eluting stents vs balloon angioplasty for the prevention of recurrence in patients with coronary in-stent restenosis: A randomised controlled trial. JAMA 2005; 293: 165-71.

[31]    Holmes DR, Teirstein P, Selter L *et al.* Sirolimus eluting stent vs vascular brachytherapy for intent restenosis with bare metal stents. The SIRS randomised trial. JAMA 2006; 295: 1264-73.

[32]   Stone GW, Ellis SG, O'Shaughnessy CD *et al.* Paclitaxel eluting stents vs vascular brachytherapy for in-stent restenosis with bare metal stents. The TAXUS V ISR randomised trial. JAMA 2006; 295: 1253-63.

[33]   Ellis SG, O'Shaughnessy CD, Martin SD. Two year clinical outcomes after paclitaxel eluting stent or brachytherapy treatment for bare metal stent restenosis: The TAXUS V ISR trial. Eur Heart J 2008; 29: 1625-34.

[34]   Westedt U, Barbu-Tudoran L, Schaper AK *et al.* Deposition of nanoparticles in the arterial vessel by porous balloon catheters: localisation by confocal laser scanning microscopy and transmission electron microscopy. AAPS Pharm Sc 2002; 4: E41.

[35]   Dannenberg HD, Fishbein I, Gao J *et al.* Macrophage depletion by clodronate containing liposomes reduces neointimal formation after balloon injury in rats and rabbits. Circulation 2002; 106: 599-605.

[36]   Dannenberg HD, Fishbein I, Epstein H *et al.* Systemic depletion of macrophages by liposomal bisphosphonate reduces neointimal formation following balloon injury in rat carotid artery. J Cardiovasc Pharmacol 2003; 42: 671-9.

[37]   Joner M, Morimoto K, Kasukawa H *et al.* site specific targeting of nanoparticle prednisolone reduces in-stent restenosis in a rabbit model of established atheroma. Arterioscler Thromb Vasc Biol 2008; 28: 1960-6.

[38]   Klugherz BD, Meneveau N, Chen W *et al.* Sustained intramural retention and regional redistribution following local vascular delivery of polylactic-coglycolic acid and liposomal nanoparticle formulations containing probucol. J Cardiovasc Pharmacol Ther 1999; 4: 167-174.

[39]   Cohen-Sela E, Rosenzweig O, Gao J *et al.* Alendronate loaded nanoparticles deplete monocytes and attenuate restenosis. J Control Release 2006; 113: 23-30.

[40]   Cohen-Sela E, Dangoor D, Epstein H *et al.* Nanospheres of a bisphosphonate attenuate intimal hyperplasia. J Nanosci Nanotechnol 2006; 6: 3226-34.

[41]   Mei L, Song CX, Jin X *et al.* Surface modified paclitaxel loaded nanoparticles as local delivery system for the prevention of vessel restenosis. Yao Xue Xue Bao 2007; 42: 81-6.

[42]   Westedt U, Kalinowski M, Wittmer M *et al.* Poly (vinyl alcohol) –graft-poly (lactide-co-glycolide) nanoparticles for local delivery of paclitaxel for restenosis treatment. J Control Release 2007; 119: 41-51.

[43]   Deshpande D, Devalapally H, Amiji M. Enhancement in anti-proliferative effects of paclitaxel in aortic smooth muscle cells upon co-administration with ceramide using biodegradable polymeric nanoparticles. Pharm Res 2008; 25: 1936-47.

[44]   Varshosaz J, Soheili M. Production and *in vitro* characterization of lisinopril loaded nanoparticles for the treatment of restenosis in-stented coronary arteries. J Microencapsul 2008; 25: 478-86.

[45]   Cohen-Sela E, Teitlboim S, Chorny M *et al.* Single and double emulsion manufacturing techniques of an amphiphilic drug in PLGA nanoparticles: Formulations of mithramycin and bioactivity. J Phrm Sci 2009; 98: 452-62.

[46]   Zou W, Cao G, Xi Y, Zhang N. New approach for local delivery of rapamycin by bioadhesive PLGA-carbopol nanoparticles. Drug Deliv 2009: 16: 15-23.

[47]   Zweers ML, Engbers GH, Grijpma DWb *et al.* J. Release of anti-restenosis drugs from poly (ethylene oxide)-poly(DL-lactic-co-glycolic acid) nanoparticles. J Control Release 2006; 114: 317-24.

[48]   Luderer F, Lobler M, Rohm HW *et al.* Biodegradable sirolimus loaded poly(lactide) nanoparticles as drug delivery systems for the prevention of in-stent restenosis in coronary stent application. J Biomater Appl 2011; 25(8): 851-75.

[49]   Nakano K, Egashira K, Masuda S *et al.* Formulation of nanoparticle eluting stents by a cationic electrodesposition coating technology: efficient nano-drug delivery *via* bioabsorbable polymeric nanoparticle eluting stents in porcine coronary arteries. JACC Cardiovasc Interv 2009; 2: 277-83.

[50]   Fishbein I, Chorny M, Robinovich L *et al.* Nanoparticulate delivery system of a tyrphostin for the treatment of restenosis. J Contol Release 200; 65: 221-9.

[51]   Banai S, Wolf Y, Golomb G *et al.* PDGF receptor tyrosine kinase blocker AG1295 selectively attenuates smooth muscle call growth *in vitro* and reduced neointimal formation after balloon angioplasty in swine. Circulation 1998; 97: 1960-9.

[52]   Banai S, Chorny M, Gertz SD *et al.* Locally delivered nanoencapsulated tyrphostin (AGL2043) reduces neointimal formation in balloon injured rat carotid and stented porcine coronary arteries. Biomaterials 2005; 26: 451-61.

[53]   Nguyen KT, Shukla KP, Moctezuma M *et al.* Studies of the cellular uptake of hydrogel nanospheres and microspheres by phagocytes, vascular endothelial cells and smooth muscle cells. J Biomed Mater Res A 2009; 88: 1022-30.

[54]   Reddy MK, Vasir JK, Sahoo SK *et al.* inhibition of apoptosis through localised delivery of rapamycin loaded nanoparticles prevented neointimal hyperplasia and reendothelialized injured artery. Circ Cardiovasc Interv 2008; 1: 209-16.

[55]   Kolodgie FD, John M, Khurana C *et al.* Sustained reduction of in-stent neointimal growth with the use of a novel systemic nanoparticle paclitaxel. Circulation 2002; 106: 1195-8.

[56]   Lanza GM, Yu X, Winter PM *et al.* Targeted antiproliferative drug delivery to vascular smooth muscle cells with magnetic resonance imaging nanoparticle contrast agent: implications for rational therapy of restenosis. Circulation 2002; 106: 2842-7.

[57]   Cyrus T, Zhang H, Allen JS *et al.* Intramural delivery of rapamycin with alphavbeta3-targeted paramagnetic nanoparticles inhibits stenosis after balloon injury. Arteriosler Thromb Vasc Biol 2008; 28: 820-6.

[58]   Chorny M, Fishbein I, Yellen BB *et al.* Targeting stents with local delivery of paclitaxel loaded magnetic nanoparticles using uniform fields. Proc Natl Acad Sci USA 2010; 107(18): 8346-51.

[59]   Margolis J, McDonald J, Heuser R *et al.* Systemic nanoparticle paclitaxel (nab-paclitaxel) for in-stent restenosis-I (SNAPIST-I): a first in human safety and dose finding study. Clin Cardiol 2007; 30: 165-70.

# CHAPTER 5

# Applications of Nanotechnology in Imaging and Therapy of Cancer

## Xiang-Hong Peng[‡], Debatosh Majumdar[‡] and Dong M. Shin[*]

*Department of Hematology and Medical Oncology, Winship Cancer Institute, Emory University School of Medicine, Atlanta, GA, 30322, USA*

**Abstract:** Cancer nanotechnology is a multidisciplinary research area in which science, engineering and medicine embrace each other to actualize the application of nanotechnology in molecular imaging, molecular diagnosis, and targeted therapy of cancer. Nanometer-sized particles such as semiconductor quantum dots, iron oxide nanocrystals and colloidal gold have unique optical, magnetic and physical properties. The surface of these nanoparticles can be conveniently functionalized to conjugate monoclonal antibodies, peptides, aptamers and small molecules, and the resulting nanoscale conjugates can be used for targeted cancer imaging and targeted cancer therapy. Recent advances in nanotechnology have led to the development of various new agents, such as nanoparticles useful for cancer imaging in multiple modalities, and nanoscale theragnostic agents suitable for simultaneous use in cancer diagnosis as well as cancer therapy.

**Keywords:** Theragnostics; nanoparticles; cancer; imaging; pharmacodynamics; uptake; tumour angiogenesis; quantum dots; tumour therapy.

## INTRODUCTION

Tumor imaging plays a crucial role in identifying solid tumors, detecting tumor recurrence and monitoring the therapeutic effects of treatment. However, conventional tumor imaging modalities such as X-ray, Computed Tomography (CT), Magnetic Resonance Imaging (MRI) and Positron Emission Tomography (PET) have limited capabilities in providing specificity and functional information regarding the cancer [1, 2], especially for the early-stage detection of tumor. For example, it is still difficult to detect a tumor of less than 5 mm diameter by any conventional imaging technique. Molecular imaging can provide more detailed functional and physiological information of a disease in specific tissues and organs with high sensitivity and specificity [3-8]. Recent advances in tumor biology, cellular and molecular biology, imaging techniques and nanotechnology have led to the development of nanoparticle probes for molecular and cellular imaging and integrated nanodevices for cancer detection and screening. For example, Iron Oxide (IO) nanoparticle-labeled human colon carcinoma cells can be detected *in vitro* by using MR imaging techniques [9-11].

As this field has developed, new agents have been designed and synthesized that can be used in imaging as well as therapy. This type of bimodal agent can diagnose and treat cancer at the same time. In this regard, a new approach known as 'theragnostics' has evolved as a treatment strategy based on the fusion between diagnostics and therapeutics [12-16]. In this approach, the results of diagnostic tests are used to select the appropriate therapy. Thus, diagnostics and therapy work hand in hand. Tremendous effort is being made to develop agents that can potentially be used for targeted cancer imaging as well as therapy.

Usually, solid tumors posses some unique properties such as hypervasculature, leaky vasculature and poor lymphatic drainage, nanoparticles having a size larger than 100nm can circulate in the vasculature for a longer time (higher retention) and extravasate into the tumor interstitium through the leaky vasculature. This passive targeting of nanoparticles is accomplished through increased permeability of the tumor vasculature. This phenomenon is commonly known as "enhanced permeability and retention" (EPR) effect

---

**\*Address correspondence to Dong M. Shin:** Winship Cancer Institute, Room 3090, 1365-C Clifton Road, Atlanta, GA 30322 (USA); Tel: 404-778-5990; Fax: 404-778-5520; E-mail: dmshin@emory.edu
‡Equal contribution.

(Fig. **1**) [17]. Non-targeted nanoparticles having a size smaller than 100nm can return into the blood capillaries from the tumor interstitium by simple diffusion. Hence, non-targeted small size nanoparticles (smaller than 100nm) are unable to reach the tumor site in sufficient quantities, and thus cannot produce signal strong enough for detection [18].

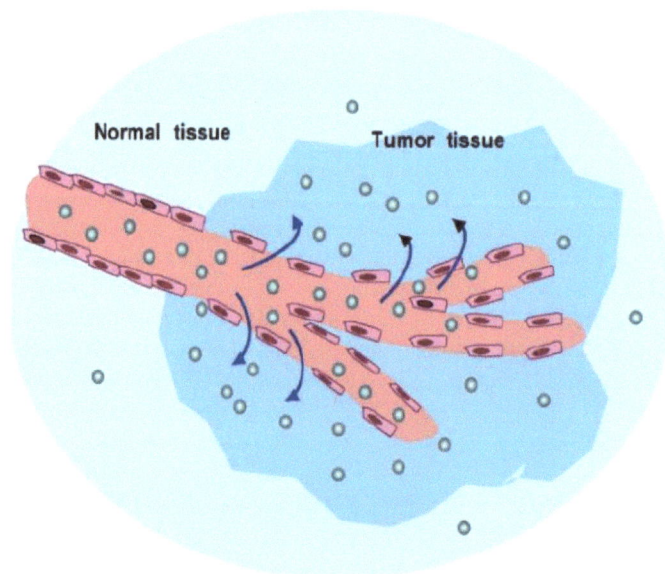

**Figure 1:** Passive tumor targeting with nanoparticle. Non-targeted nanoparticles accumulate passively in solid tumor tissue by the enhanced permeability and retention effect (EPR). The hyperpermeable angiogenic tumor vasculature allows preferential extravasation of circulating nanoparticles. Reproduced with permission from Wang X, Yang L, Chen ZG, Shin DM. Application of nanotechnology in cancer therapy and imaging. CA Cancer J Clin. 2008, 58 (2): 97-110.

In order to improve the specificity and sensitivity of cancer detection, it is urgently needed to develop target-specific nanoparticles as imaging probes. This has been successfully accomplished by utilizing cancer cell-specific biomarkers and novel nanoparticles [8, 19-21]. Recently, target specific optical, radioactive, and magnetic probes have been developed and their feasibility for application in cancer imaging was examined *in vivo* [22-25]. The results of these studies have shown that tumor-targeted imaging probes can increase the local concentration of imaging probes in tumors while reducing their uptake in normal tissues. To date, various novel nanoparticles have been developed for tumor imaging.

At the beginning of this chapter, we focus on recent advances in the development of novel nanoparticles, such as quantum dots (QDs), gadolinium nanoparticles, gold and superparamagnetic iron oxide (IO) nanoparticles as imaging probes. In the later part, we discuss the use of nanoparticles in multimodality imaging. Finally, we discuss the field of theragnostics–the fusion between therapy and diagnostics.

## 1. QUANTUM DOT (QD) NANOPARTICLES

Semiconductor quantum dots (QDs) have unique optical and electronic properties when compared with commonly used organic dyes. For example, QDs have extremely high levels of brightness (10-100 times brighter) and excellent photostability (100-1000 times more stable against photobleaching), large absorption co-efficient across a wide spectral range, a sharp fluorescence emission spectra, and their fluorescence emission can be tuned from visible to infrared wavelength by varying their composition and size (Fig. **2**) [26]. Moreover multicolor QDs probes can be used for simultaneous imaging and tracking of multiple tumor markers, further increasing both the specificity and sensitivity of cancer detection. Such structurally unique QD probes have a versatile structure suitable for conjugation to therapeutic and imaging agents.

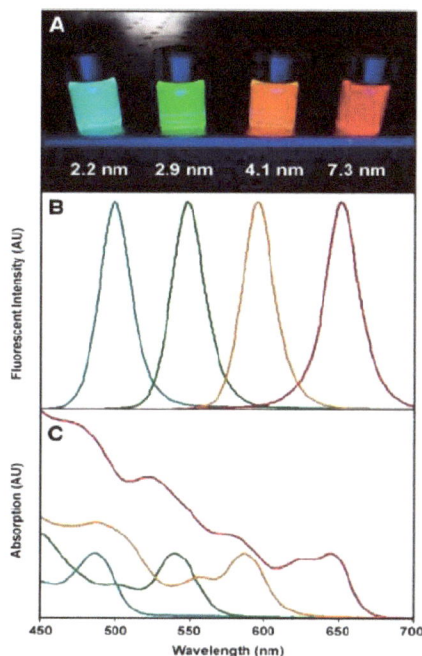

**Figure 2:** Size-dependent optical properties of cadmium selenide QDs dispersed in chloroform, illustrating quantum confinement and size-tunable fluorescence emission. (A) Fluorescence image of four vials of monodisperse QDs with sizes ranging from 2.2 nm to 7.3 nm in diameter. (B) Fluorescence spectra of the same four QD samples. Narrow emission bands (23–26 nm full-width at half-maximum, WHM) indicate narrow particle size distributions. (C) Absorption spectra of the same four QD samples. Reproduced with permission from Smith AM, Ruan G, Rhyner MN, Nie S. Engineering luminescent quantum dots for *in vivo* molecular and cellular imaging. Ann Biomed Engineering 2006, 34 (1): 3-14.

A new class of semiconductor QD conjugate **1** has been developed for *in vivo* cancer targeting and fluorescence imaging of prostate cancer (Figs. **3** and **4**) [27]. This QD conjugate consists of an amphiphilic triblock copolymer coating on the QD surface used for *in vivo* protection, targeting ligands to recognize the tumor antigen prostate specific membrane antigen (PSMA), and PEG molecules to improve biocompatibility, bioavailability, and circulation. PSMA is overexpressed in prostate cancer lines, such as C4-2 cells, but is absent in PC-3 cells. The QD-PSMA antibody conjugate probe **1** was injected into the tail vein of a tumor-bearing mouse harboring a C4-2 tumor xenograft and a control mouse (having no tumor) [27]. *In vivo* imaging, immunocytochemical and histological studies confirmed that the fluorescence signals of the QD-PSMA antibody conjugate came from the tumor containing C4-2 cells due to active targeting (Fig. **5**), whereas, in the control mouse lacking tumor, the QD-PSMA antibody conjugate probe **1** did not induce any QD signal [27]. Further studies using near-infrared emitting QDs are being explored to see if this can improve tissue penetration depth and imaging sensitivity.

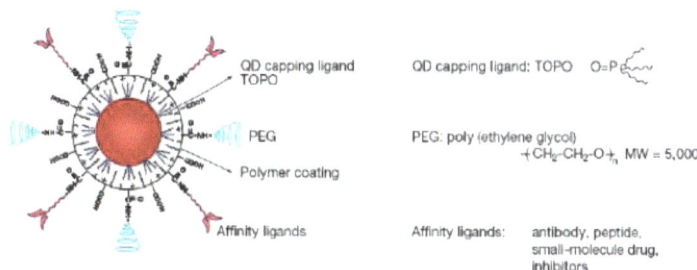

**Figure 3:** Schematic illustration of biconjugated QDs for *in vivo* cancer targeting and imaging. Reproduced with permission from Gao X, Cui Y, Levenson RM, Chung LW, Nie S. *In vivo* cancer targeting and imaging with semiconductor quantum dots. Nat Biotechnol 2004; 22 (8): 969-76.

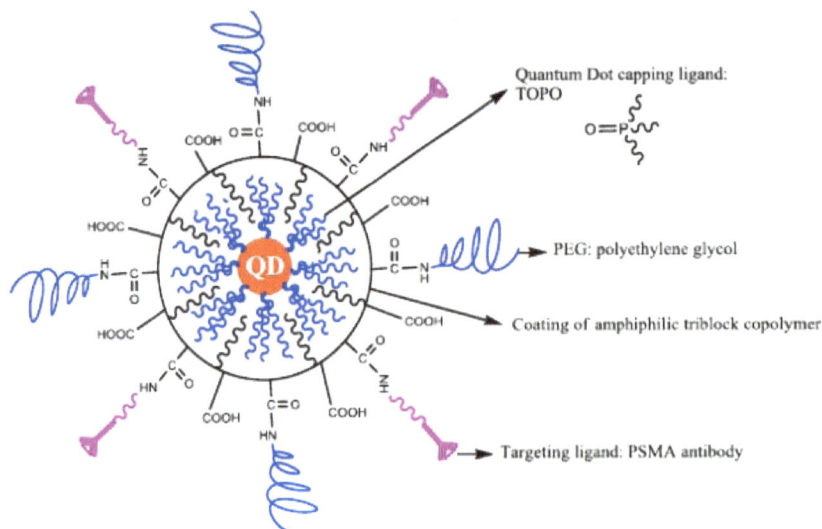

**Figure 4:** QD-PSMA antibody conjugate nanoparticle. This figure is drawn based on the the article, *Nature Biotechnol* 2004; 22 (8): 969-76.

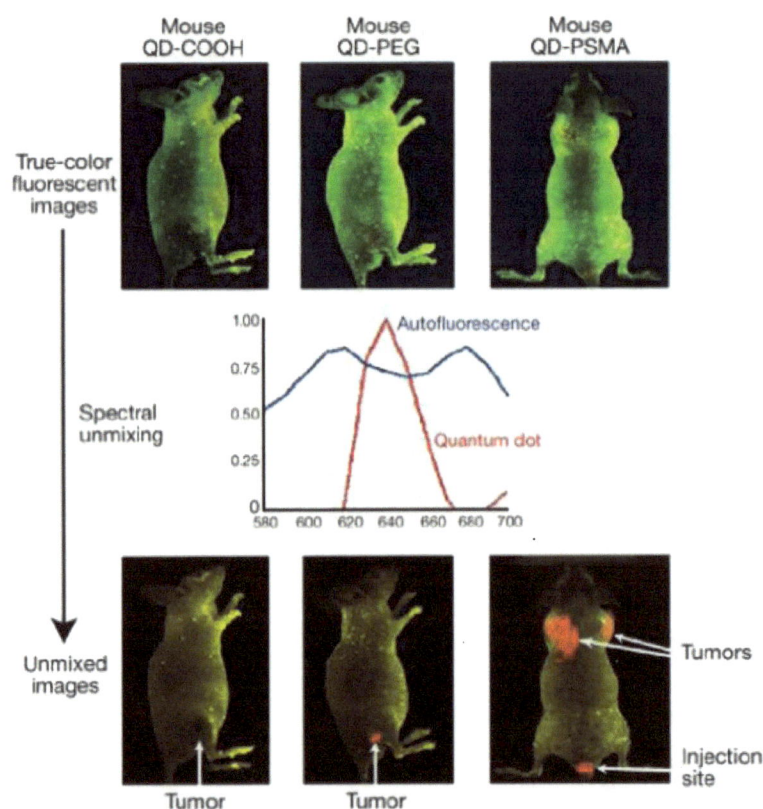

**Figure 5:** *In vivo* fluorescence images of tumor-bearing mice using QD probes with three different surface modifications: carboxylic acid groups (left), PEG groups (middle) and PEG-PSMA Ab conjugates (right). For each surface modification, a color image (top), two fluorescence spectra from QD and animal skin (middle) and a spectrally resolved image (bottom) were obtained from the live mouse models bearing C4-2 human prostate tumors of similar sizes (0.5−1.0 cm in diameter). The site of QD injection was observed as a red spot on the mouse tail. The spectral feature at ~700 nm (red curve, middle panel) was an artifact caused by mathematical fitting of the original QD spectrum, which has little or no effect on background removal. Reproduced with permission from Gao X, Cui Y, Levenson RM, Chung LW, Nie S. *In vivo* cancer targeting and imaging with semiconductor quantum dots. Nature Biotechnol 2004; 22 (8): 969-76.

Recently, our group used secondary antibody-conjugated QDs with two different emission wavelengths (QD605 and QD565) to analyze the expression of epidermal growth factor receptor (EGFR) and E-cadherin (E-cad) simultaneously in the same head and neck cancer cells [28]. This study showed that QD-immunocytochemistry (QD-ICC)- based technology can not only quantify basal levels of multiple biomarkers but also track the localization of the biomarkers upon biostimulation (Fig. **6**) [28].

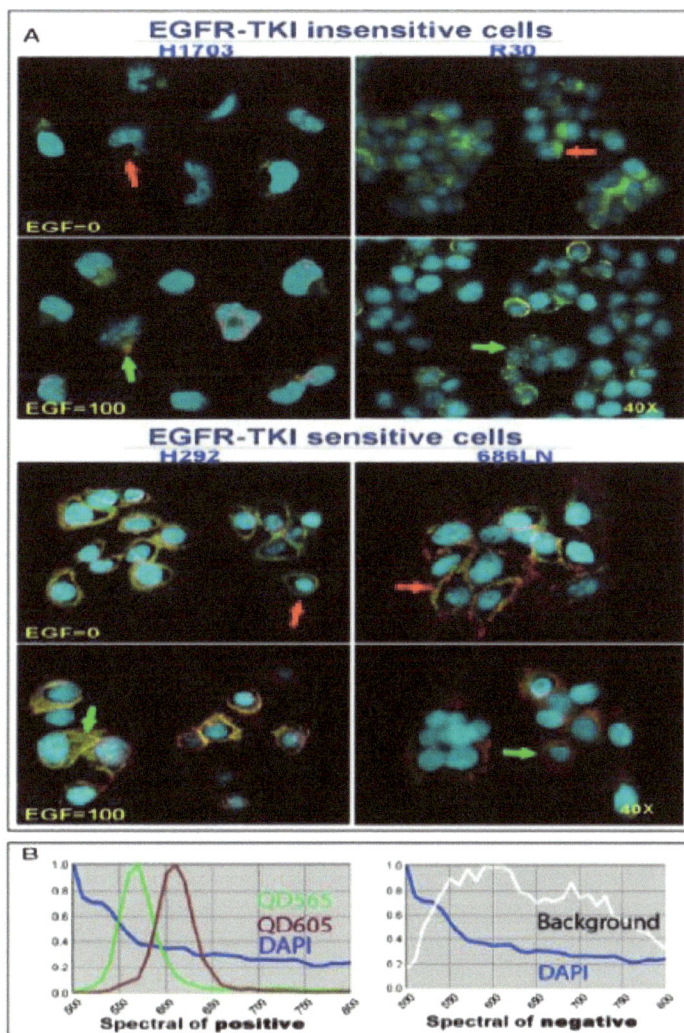

**Figure 6:** (A) Colocalization of EGFR (QD605, red) and E-cad (QD565, green) in the absence or presence of EGF. EGFR and E-cad expression levels were imaged by an Olympus microscope IX71 with a CRi spectral imaging and quantifying system. Yellow indicates colocalization of EGFR (red) and E-cad (green). (B) Spectral image of positive staining (including QD 565, QD 605, and DAPI signals) and negative staining (only background and DAPI signals). Reproduced with permission from Huang DH, Su L, Peng XH, Zhang H, Khuri FR, Shin DM, *et al.* Quantum dot-based quantification revealed differences in subcellular localization of EGFR and E-cadherin between EGFR-TKI sensitive and insensitive cancer cells. Nanotechnol 2009; 20 (22): 225102.

Blood vessels of individual organs, tissues, and tumors express molecular markers that are characteristic of their individual vasculature. Peptides can specifically recognize these vascular markers. For example, the peptide KDEPQRRSARLSAKPAPPKPE-PKPKKAPAKK (F3) preferentially binds to blood vessels and tumor cells in various tumors, and CGNKRTRGC (LyP-1) peptide recognizes lymphatic vessels and tumor cells in certain tumors. *In vivo* studies by Akerman *et al.* demonstrated that F3 peptide-QD conjugates could co-localize with the blood vessel in tumor tissue, whereas LyP-1 peptide-QD conjugates accumulated in the tumor tissue, but did not co-localize with the blood vessel marker [29].

Various research groups have reported the use of QD-peptide/antibody conjugates for specific targeting of liver, breast and pancreatic cancer, and glioblastoma tumors *in vivo* [25, 30-36]. Moreover, QDs can be used to study the complex anatomy and pathophysiology of cancer in animal models. Jain *et al.* has shown the synergy between QDs and multiphoton intravital microscopy in differentiating tumor vessels from perivascular cells and matrix, investigated the ability of microparticles to access the tumor, and monitored the trafficking of precursor cells [37].

One disadvantage of using QDs for *in vivo* imaging is their limited depth penetration (<1 cm). To overcome this obstacle, QDs have recently been developed that can produce near infrared (NIR) signals [32, 38]. NIR light penetrates much deeper into tissues compared to the visible fluorescence, allowing detection of signals in the deep tissue of animals. More importantly the emission of NIR QDs is well beyond the spectral range of tissue autofluorescence, and provides image with a high signal-to-background ratio. The real time detection of QD NIR signals in sentinel lymph nodes of large animals has successfully been demonstrated [39].

Hence, QDs are excellent optical imaging nanoprobes for evaluating the specificity of tumor targeting ligands *in vitro* in tumor cells and *in vivo* in animal tumor models. In addition, QD-based optical imaging could be combined with other imaging modalities such as MRI and positron emission tomography (PET) for quantitative analysis of the biodistribution and targeting efficacy [33].

However, commonly used QDs are composed of cadmium. Cadmium is potentially hazardous and toxic to living cells and humans. There are other factors such as size, charge, concentration, surface coating of QDs, and oxidative, photolytic and mechanical stability that can affect the toxicity of QDs [40]. Work is in progress to reduce the toxicity of QDs and to understand the complete course of QDs *in vivo*.

## 2. GOLD NANOPARTICLES

The unique physical, chemical and biological properties of gold nanoparticles make them suitable for use as CT contrast agents, detection of surface-enhanced Raman scattering (SERS) and probes for reflectance imaging.

There are many advantages in using gold nanoparticles in targeted tumor imaging, such as 1) their small size, 2) strong absorbing and scattering properties, 3) high contrast (gold absorbs ~3 times more than an equal weight of iodine at 20 KeV), 3) no photobleaching or blinking, 4) low osmolality, even at high concentrations, 5) low viscosity, similar to water, 6) biocompatibility and considerably less cytotoxicity *in vivo*, 7) easy preparation and bioconjugation, 8) permeates angiogenic endothelium of tumors, 9) longer blood residence time than commonly used iodine agents (iopromide, iohexol etc), 10) can image using standard MicroCT, clinical CT, planar X-ray, mammography [41].

Computed Tomography (CT) is widely used for *in vivo* tumor imaging. Iodinated molecules such as iopromide and iohexol are the main CT contrast agents used in clinic for a long time. However iodinated molecules can cause side effects to patients having impaired renal function [42]. These effects range from mild acute symptoms such as nausea and skin irritation to organ-specific reactions such as contrast-induced nephropathy [42]. Moreover, the images generated by the non-specific iodinated agents have very short life time because of their rapid clearance by the kidneys.

Gold nanoparticles have higher absorption than iodine with less bone and tissue interference accomplishing better contrast with lower X-ray dose. Gold nanoparticles clear the blood more slowly than iodine agents, allowing longer imaging times. More importantly, surface coating on gold nanoparticles provides active functional groups amenable for conjugation of targeting peptides or antibodies making them suitable for targeted imaging.

A new class of immune-targeted gold nanoprobes has been developed for *in vivo* CT molecular imaging that can selectively and sensitively target specific antigens with sensitively. Gold nanorods were conjugated with UM-A9 antibodies to target the A9 antigens which are overexpressed in Squamous Cell Carcinoma of

Head and Neck (HNSCC) [22]. Accumulation of high concentration of gold nanoparticles on the targeted cancer cells generated a strong selective X-ray attenuation that was distinct from the attenuation obtained by normal cells or non-targeted cancer cells [22].

The EGFR signaling pathways play crucial role in the regulation of cell proliferation, survival and differentiation. Overexpression of EGFR is one of the most common molecular alterations in many tumors, thus making it an ideal tumor target in designing nanoparticle-based imaging probe or drug delivery agent [43-47].

Gold nanoparticles have strong localized surface plasmon resonance that make them ideal probes for reflectance imaging (Fig. **7**). Due to the surface plasmon absorption and light scattering properties of gold nanorods, they have been used as contrast agents for molecular imaging of EGFR-expressing cancer cell using dark-field microscopy. In this regard, anti-EGFR antibody was conjugated to gold nanorods. Two malignant oral epithelial cell lines (HOC 313 clone 8 and HSC 3) and one non-malignant epithelial cell line (HaCat) were used. After incubation with the cells, the light scattering images of anti-EGFR antibody-conjugated gold nanorods have demonstrated that the anti-EGFR antibody-conjugated gold nanorods bind selectively to the surface of malignant cells with much higher binding affinity owing to the high overexpression of EGFR on the cytoplasmic membrane of the malignant cells [48].

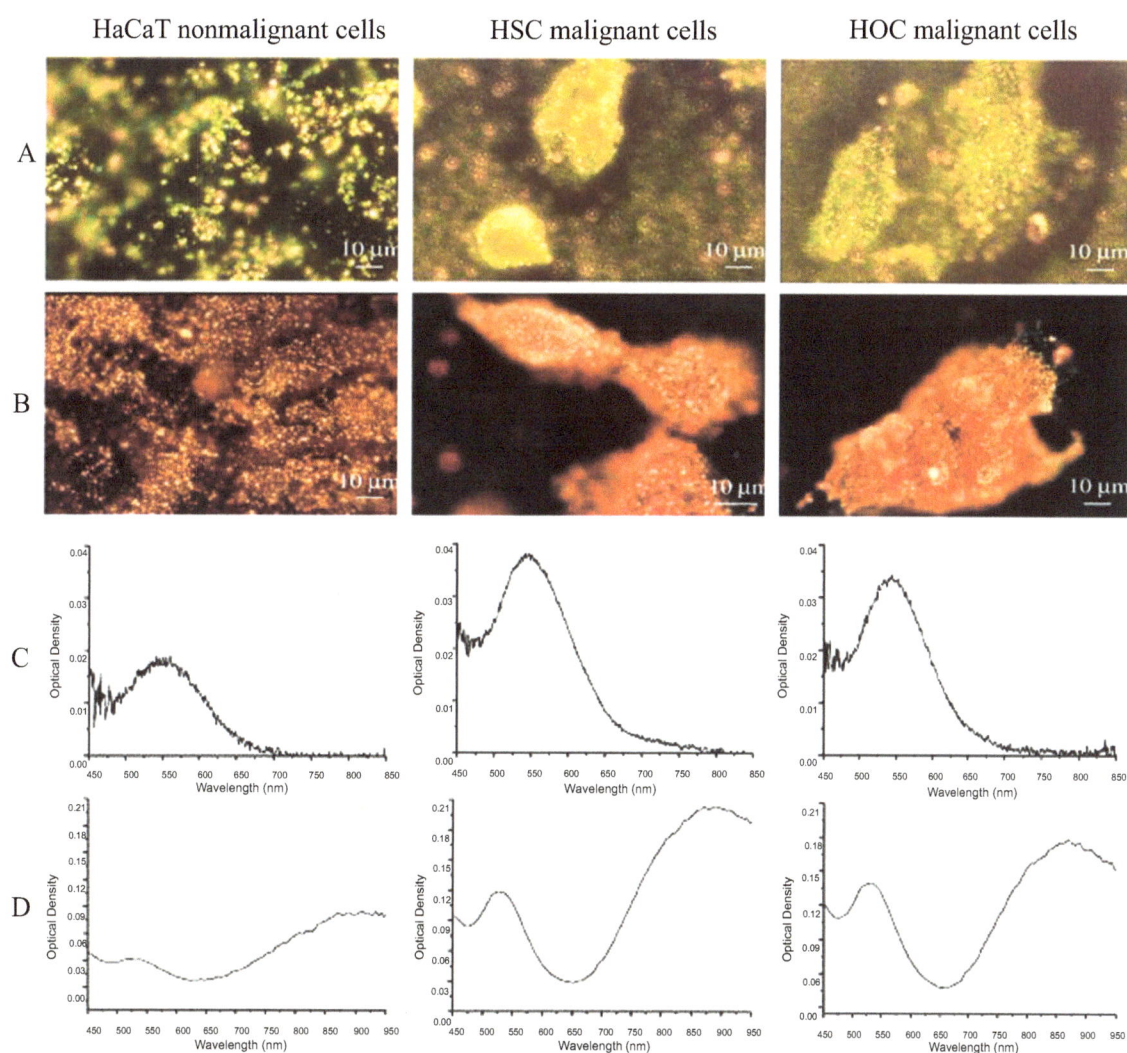

**Figure 7:** (A) Light scattering images of anti-EGFR/Au nanospheres after incubation with cells for 30 min at room temperature. (B) Light scattering images of anti-EGFR/Au nanorods after incubation with cells for 30 min at room temperature. (C) Average extinction spectra of anti-EGFR/Au nanospheres from 20 different single cells for each kind.

(D) Average extinction spectra of anti-EGFR/Au nanorods from 20 different single cells for each kind. From gold nanospheres, the green to yellow color is most dominant, corresponding to the surface plasmonic enhancement of scattering light in the visible region, and from gold nanorods, the orange to red color is most dominant, corresponding to the surface plasmonic enhancement of the longitudinal oscillation in the near-infrared region. Reproduced with permission from Huang X, El-Sayed IH, Qian W, El-Sayed MA. Cancer cell imaging and photothermal therapy in the near-infrared region by using gold nanorods. J Am Chem Soc 2006; 128 (6): 2115-20.

A new class of gold nanoparticle-based molecular specific contrast agent has been developed for vital optical imaging of precancers and cancers to probe molecules with high affinity for cellular biomarkers such as EGFR [49]. The cervical epithelial cancer cell line SiHa highly overexpresses EGFR and was used in this study. Selective binding of the EGFR antibody gold-conjugates to the SiHa cells was demonstrated using laser scanning confocal reflectance and combined confocal reflectance/transmittance imaging of labeled SiHa cells [49]. It was also shown that for the purpose of imaging, gold conjugates can be delivered topically throughout the whole epithelium. These gold contrast agents have potential applications for *in vivo* molecular imaging [49].

Gold nanoparticles have been used for detection of surface-enhanced Raman scattering (SERS) (Fig. **8**) [50]. One of the great advantages of colloidal gold nanoparticles is their ability to amplify the Raman scattering efficiencies of adsorbed molecules by as much as $10^{14}$-$10^{15}$–fold, allowing spectroscopic detection and identification of single molecules. Our research group has developed a new class of biocompatible and nontoxic nanoparticles for *in vivo* tumor targeting and detection by using pegylated gold nanoparticles and SERS tag [50]. The SERS tag-containing pegylated gold nanoparticles consist of three components: 60 nm citrate-stabilized gold nanoparticle, Raman reporter molecule, and thiolated polyethylene glycol (PEG-SH) (Fig. **8**) [50]. Due to the PEG coating, which could protect gold nanoparticles from aggregation in concentrated electrolyte solutions and organic solvents, these pegylated gold nanoparticles show minimal non-specific binding and very low cytotoxicity in intracellular delivery studies. The biocompatible polymer PEG provides diverse functional groups for efficient conjugation of biological targeting ligands. These Pegylated SERS nanoparticles are about 200 times brighter than semiconductor QDs with light emission in the same infrared window. By conjugating a single-chain anti-EGFR antibody (ScFvEGFR) to the gold nanoparticles, the EGFR on the human head and neck cancer cells was targeted with high specificity and affinity. After tail vein injection of the SERS tag-containing EGFR antibody-conjugated gold nanoparticles into nude mice bearing human head and neck cancer cells (Tu686), a significant difference was observed between the signal intensities of the SERS spectra of the targeted and nontargeted nanoparticles; whereas the SERS signal intensities from non-specific liver uptake were similar (Fig. **9**) [50]. The pegylated gold nanoparticles are also dual-modality probes for optical as well as electron microscopic imaging, offering new avenues for designing multifunctional nanoparticles for *in vivo* molecular imaging.

To date, gold nanoparticles have mainly been used for *in vitro* tumor imaging by dark field light. However, gold nanoparticles have a few limitations; for example, the optical signal of gold nanoparticles may not be strong enough for signal detection. Moreover the low tissue penetration of the optical signal further limits the application of gold nanoparticles for *in vivo* imaging.

Raman imaging combined with endoscopy has potential applications for *in vivo* biomedical imaging. The use of multiple SERS nanoparticles with different absorption wavelengths in the NIR region to simultaneously target many tumor markers will significantly improve the sensitivity and specificity of tumor imaging.

Applications of gold nanoparticles in the diagnosis of cancer have raised the issue of their probable cytotoxicity. Pernodet *et al.* reported that 14-nm gold nanoparticles could easily cross the cell membrane and accumulate into vacuoles. These nanoparticles caused abnormalities in the actin filaments and extracellular matrix constructs of dermal fibroblasts resulting in a decrease in cell proliferation, adhesion and motility [51].

Jeong *et al.* studied the toxicity and pharmacokinetics 13-nm sized PEG-coated gold nanoparticles. It was shown that these gold nanoparticles could induce acute inflammation and apoptosis in the liver of mice [52]. Moreover, TEM imaging showed that various cytoplasmic vesicles and lysosomes of liver Kupffer cells and spleen macrophages contained PEG-coated gold nanoparticles [52]. Although the temporary inflammation may be considered as a lesion caused by the host defense mechanism, the apoptosis observed is believed to be due to the toxicity of PEG-coated gold nanoparticles.

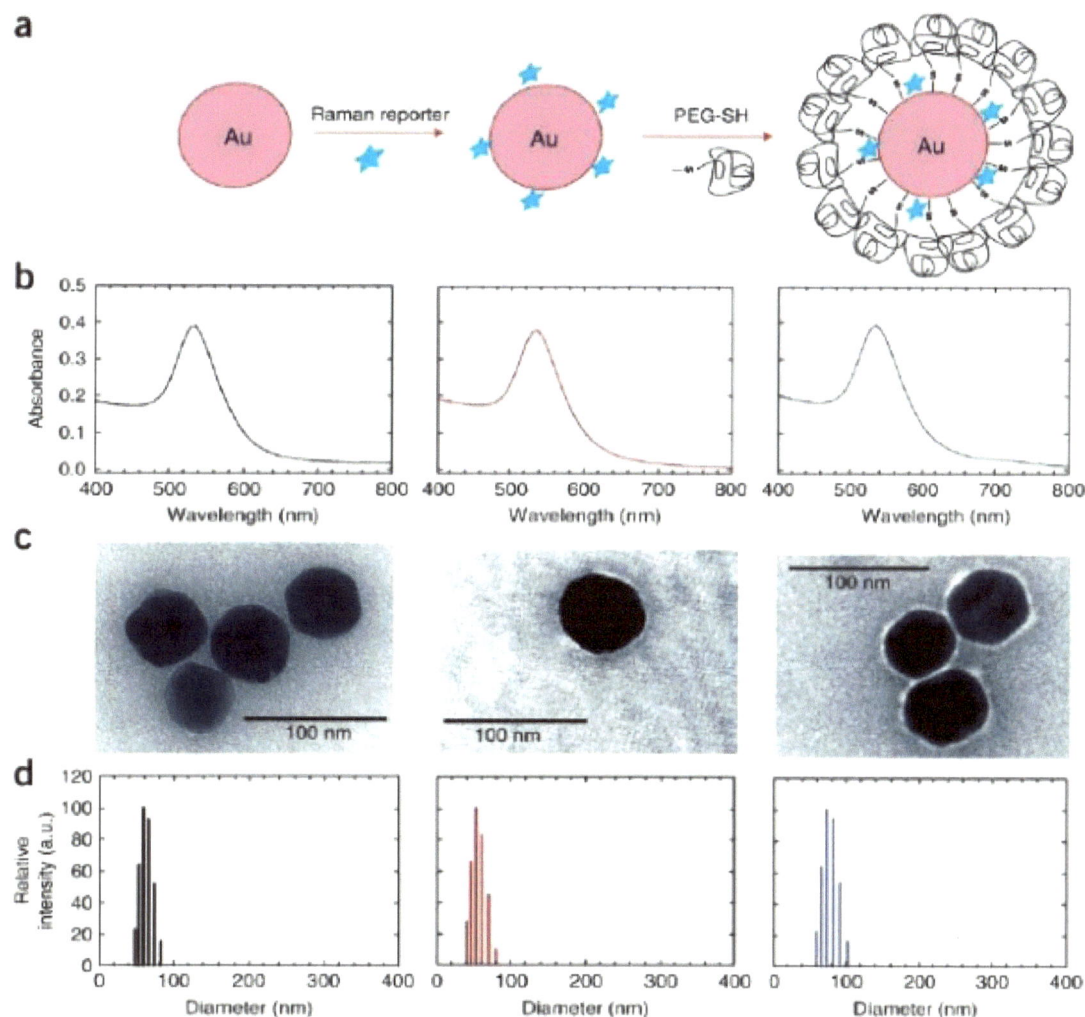

**Figure 8:** (a) Preparation and schematic structures of the original gold colloid, a particle encoded with a Raman reporter, and a particle stabilized with a layer of thiol-polyethyleneglycol (thiol-PEG). Approximately 1.4–1.5 $\times 10^4$ reporter molecules (e.g., malachite green) are adsorbed on each 60-nm gold particle, which is further stabilized with 3.0 $\times 10^4$ thiol-PEG molecules. (b) Optical absorption, (c) Transmission electron microscopy (TEM); and (d) Dynamic light scattering size data obtained from the original, Raman-encoded, and PEG-stabilized gold nanoparticles as shown in a. Reproduced with permission from Qian X, Peng XH, Ansari DO, Yin-Goen Q, Chen GZ, Shin DM, *et al. In vivo* tumor targeting and spectroscopic detection with surface-enhanced Raman nanoparticle tags. Nat Biotechnology, 2008, 26 (1): 83-90.

Recent studies from our lab have found that Pegylated gold nanoparticles can be detected in the organs of mice for over three months. Like other nanoparticles the toxicity of gold nanoparticles may depend on their particle size, shape, structure and surface chemistry. Further studies on the relationship between surface properties and biodistribution may provide insights useful in designing gold nanoparticles with greatly reduced toxicity.

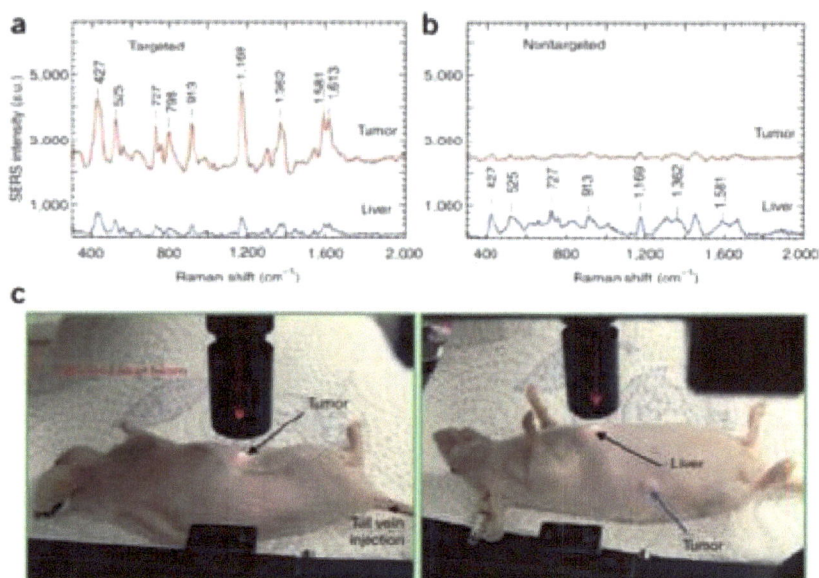

**Figure 9:** (a, b) SERS spectra obtained from the tumor and the liver locations by using targeted (a) and nontargeted (b) nanoparticles. Two nude mice bearing human head and neck squamous cell carcinoma (Tu686) xenograft tumors (3-mm diameter) received 90 μL of ScFv EGFR-conjugated SERS tags or pegylated SERS tags (460 pM). The particles were administered *via* single tail vein injection. SERS spectra were taken 5 h after injection. (c) Photographs showing a laser beam focusing on the tumor site or on the anatomical location of liver. *In vivo* SERS spectra were obtained from the tumor site (red) and the liver site (blue) with 2-s signal integration and at 785 nm excitation. The spectra were background subtracted and shifted for better visualization. The Raman reporter molecule is malachite green, with distinct spectral signatures as labeled in a and b. Laser power, 20 mW. Reproduced with permission from Qian X, Peng XH, Ansari DO, Yin-Goen Q, Chen GZ, Shin DM, *et al. In vivo* tumor targeting and spectroscopic detection with surface-enhanced Raman nanoparticle tags. Nat Biotechnology, 2008, 26 (1): 83-90.

## 3. GADOLINIUM AND MAGNETIC IRON OXIDE (IO) NANOPARTICLES

Magnetic Resonance Imaging (MRI) offers exceptionally high resolution image and can distinguish between adjacent soft tissues revealing tissue morphology and anatomical details. The excellent anatomical resolution of Magnetic Resonance (MR) images has made MRI an extremely useful technique in modern diagnostics. Since the endogenous contrast in an MR image arises mostly from differences in the relaxation times of tissue water protons, clinically used contrast agents are mainly paramagnetic chelates of gadolinium ($Gd^{3+}$) ions [53]. Their function is based on the enhancement of the tissue contrast due to the increase in the longitudinal relaxation rate of water protons ($1/T_1$). Substantial progress has been made in developing $Gd^{3+}$ ion-based MRI agents for targeted cancer imaging that can selectively target specific receptors over expressed on certain cancers.

Recent advances in nanotechnology have generated immense interest in using other nanoparticles such as iron oxide particles [8] as MRI contrast agents. However, their sensitivity is still low when used in molecular and cellular imaging [1, 54]. Magnetic contrast agents are used mainly to enhance the contrast and to amplify the signal.

### 3.1 Use of Gd Nanoparticle in Targeting $\alpha_v\beta_3$-Integrin

Integrin signaling plays an important role in tumor angiogenesis and metastasis. Integrin $\alpha_v\beta_3$ is significantly upregulated in tumor vasculature. Integrin $\alpha_v\beta_3$ has a strong binding affinity towards the arginine-glycine-aspartic acid (RGD) peptide motif. Molecular imaging of angiogenic tumor vasculature offers an excellent opportunity for early stage detection of nascent tumors or metastasis, selection of patients for appropriate antiangiogenic treatments, and evaluation of efficacy of antineovascular therapeutics [55].

Recently the *in vivo* use of MR imaging employing a $Gd^{3+}$-based $\alpha_v\beta_3$-integrin-targeted paramagnetic nanoparticle has been reported. The nanoparticles were formulated using gadolinium diethylenetriamine penta acetic acid-bisoleate (Gd-DTPA-BOA), and N-[{w-[4-(p-maleimidophenyl) butanoyl]amino} poly(ethylene glycol)2000]1,2-distearoyl-sn-glycero-3-phosphoethanol-amine covalently conjugated to the $\alpha_v\beta_3$-integrin peptidomimetic antagonist [20, 56, 57]. The particle has an average diameter of 273 nm. Vx-2 carcinoma tumor fragment was implanted in a Vx-2 rabbit tumor model. $Gd^{3+}$-based $\alpha_v\beta_3$-integrin-targeted paramagnetic nanoparticles were administered into the rabbits by intravenous injection [20, 56, 57]. A clinically relevant MRI scanner of field strength 1.5 T was used. $T_1$-weighted MR images of the Vx-2 rabbit tumors receiving the $Gd^{3+}$-based $\alpha_v\beta_3$-integrin-targeted paramagnetic nanoparticles showed a significant increase in MRI contrast, mainly along the tumor periphery [20, 56, 57]. Histological and immunocytochemical analysis of the Vx-2 tumors also supported that angiogenesis was predominantly distributed along the tumor periphery [20, 56, 57]. Nanoparticle-targeted contrast signal enhancement was specific to the sites of tumor angiogenesis and the adjoining vasculature, and to the tissue stimulated by the developing tumor.

In comparison to the clinically used $T_1$ MRI contrast agent Gd-DTPA, IO nanoparticles have many advantages as $T_2$ MRI contrast agents, such as, 1) high contrast, 2) altered magnetic properties, 3) long blood retention time, and 4) biodegradability and low toxicity [1, 58-61]. The magnetism of IO nanoparticles and their effect on MR imaging depend significantly on their morphology, crystal structure, size and uniformity. Different sizes of IO nanoparticles including superparamagnetic iron oxide (SPIO 60-150 nm), ultrasmall superparamagnetic iron oxide (USSPIO 10-50 nm) and monocrystalline iron oxide (MIO) posses different magnetic properties and functions for various applications [62].

Tumor target-specific antibodies, peptides and small molecules are usually conjugated to the surface of IO nanoparticles to selectively target the specific tumor type. To date, antibody-based targeted IO nanoparticles have been widely studied for *in vitro* and *in vivo* tumor imaging [63], and results have demonstrated high target specificity and generated strong signal to background ratio.

### 3.2 Use of Iron Oxide (IO) Nanoparticle in Targeting of $\alpha_v\beta_3$-Integrin

A novel approach to detect *in vivo* tumor angiogenesis by $\alpha_v\beta_3$-targeted magnetic resonance imaging using LM609 antibody-conjugated paramagnetic liposomes (ACPLs) [64] has been reported (Fig. **10**, schematic representation **3**). A new class of paramagnetic polymerized liposome particle (PPL) **2** [65] has been developed, which has lanthanide ion chelates as head groups, and can be visualized using MRI (Fig. **10**, schematic representation **2**). Biotin has been incorporated into the particle for use as a marker for histochemical studies. These new particles (PPLs) **2** have been conjugated to LM609 to obtain the LM609 ACPLs (Fig. **10**, schematic representation **3**) [64]. *In vivo* imaging studies were performed using a rabbit model of squamous cell carcinoma (V2). Rabbits having tumors of about 3-4 cm were injected intravenously with either LM609 ACPLs or control ACPLs and imaged on a 1.5 T clinical scanner [64]. LM609 ACPLs conferred $T_1$-weighted magnetic resonance signal enhancement of rabbit tumors [64]. Due to the high $\alpha_v\beta_3$ expression level on vascular endothelium, high signal intensity enhancement was observed at the tumor periphery, but not within the tumor. The mean signal intensity enhancement data at 24 h post-contrast administration using $\alpha_v\beta_3$-targeted ACPLs was more than twofold higher than that of isotype matched controls, and more than tenfold greater than that of streptavidin-biotin-labeled paramagnetic liposomes [64]. This study demonstrates that the endothelial tumor angiogenesis marker $\alpha_v\beta_3$ can be visualized *in vivo* using this novel targeted multivalent MRI agent, LM609 ACPLs. This approach enables visualization of tumor angiogenesis non-invasively and quantitatively, with high molecular specificity and high anatomic resolution of MRI.

Imaging of tumor angiogenesis will play a crucial role in modern-day cancer management. Moreover, the probes developed for imaging of tumor angiogenesis can have potential applications in other angiogenesis-related diseases such as atherosclerosis, myocardial infarction, stroke, chronic inflammation, and peripheral artery disease.

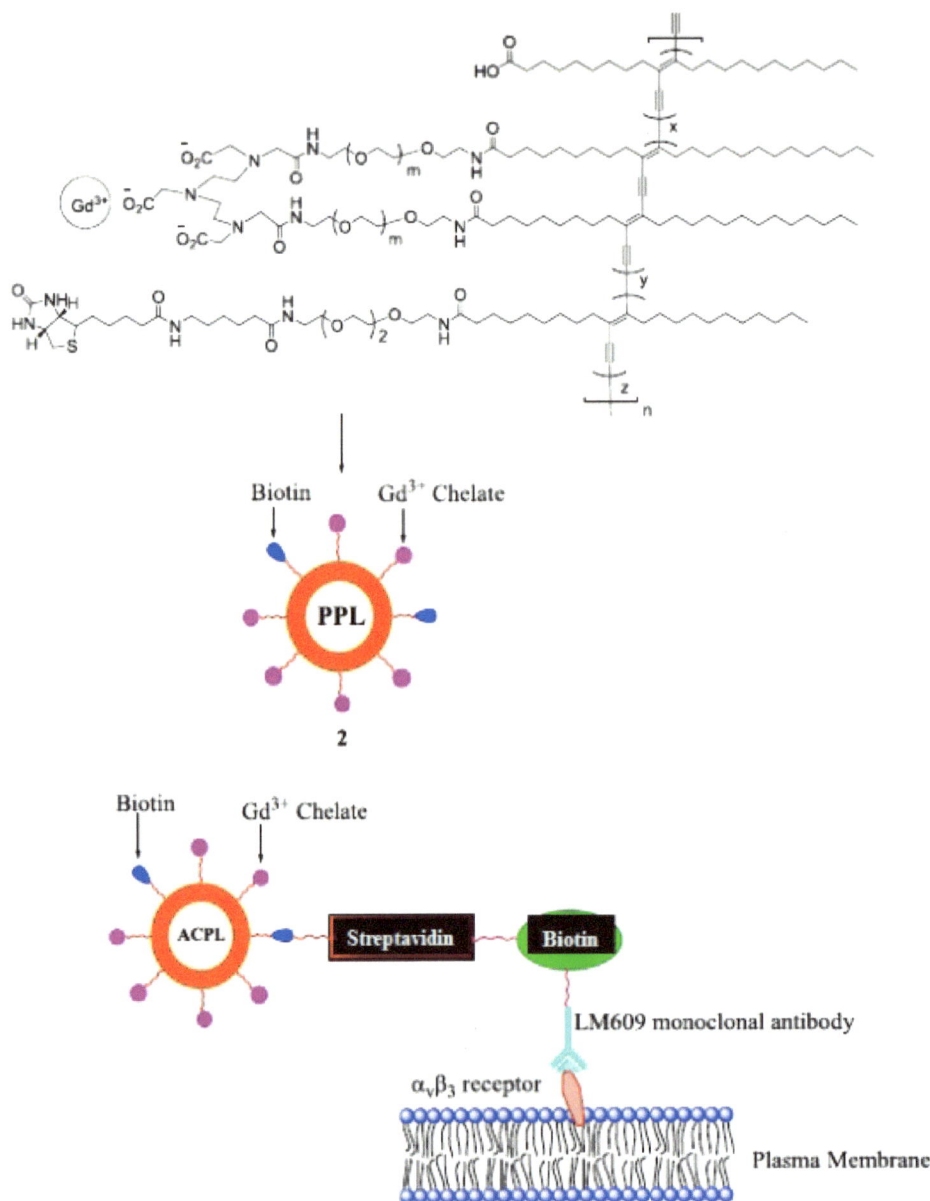

**Figure 10:** Schematic representation of formation of paramagnetic polymerized liposome (PPL) **2**, and $\alpha_v\beta_3$ targeted Magnetic Resonance Imaging (MRI) using antibody-conjugated paramagnetic liposomes (ACPLs) **3**. This figure is drawn based on the articles Nat Med, 1998, May, 4 (5): 623-6, and J Magn Reson Imaging 1995; Nov-Dec, 5(6): 719-24.

### 3.3 Use of Iron Oxide Nanoparticles in Targeting of Her-2/neu Receptor

Human epidermal growth factor receptor 2 (Her-2/*neu* receptor) is a cell membrane surface-bound receptor tyrosine kinase. It is involved in the signal transduction pathways leading to cell growth and differentiation. Her-2/*neu* receptor is particularly overexpressed in breast cancer, also in ovarian, stomach, and uterine cancer.

MR molecular imaging has been performed using a Superparamagnetic Iron Oxide (SPIO) nanoparticle contrast agent targeted to the Her-2/*neu* receptor overexpressed in breast cancer cells [66]. SPIO nanoparticle coated with a polysaccharide layer having a diameter of 50 nm was conjugated to streptavidin, to give SPIO-streptavidin nanoparticle, and used as a targeted $T_2$ contrast agent (Fig. **11**) [66]. The humanized monoclonal anti Her-2/neu antibody Herceptin was used to recognize the extracellular domain of the Her-2/*neu* receptor. Herceptin was biotinylated so that the SPIO-streptavidin nanoparticle could bind Herceptin through the biotin-streptavidin interaction [66]. Three human breast cancer cell lines AU-565, MCF-7, and MDA-MB-231 were used, which express high, moderate, and low levels of Her-2/*neu* receptor, respectively. It was observed that the $T_2$ contrast was most prominent in AU-565 cells and less prominent in MCF-7 and MDA-MB-231 cells [66]. A comparative study of the change in $T_2$ relaxation rate, $\Delta(1/T_2)$ versus the concentration of receptor sites demonstrated a positive trend in $(1/T_2)$ relaxation rate with increased concentration of contrast agent, and the $(1/T_2)$ relaxation rate increased linearly with the increasing number of SPIO binding (receptor) sites [66]. The linear dependence of $(1/T_2)$ relaxation rate on SPIO concentration is a great advantage of $T_2$ MRI for the detection of cancer cells using SPIO nanoparticles targeted to cell surface receptors. The typical expression level of Her-2/neu receptors in clinical cases of breast cancer is in the range of $1 \times 10^5$ to $4.5 \times 10^6$ per cell. This SPIO nanoparticle contrast agent demonstrated high sensitivity with the lower limit of detection in the range of $5 \times 10^4$ receptors per cell.

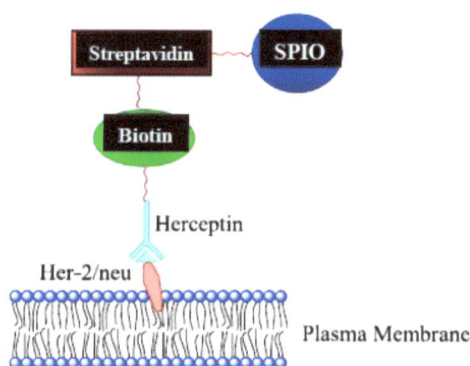

**Figure 11:** Her-2/neu receptors are targeted using SPIO nanoparticles. This figure is drawn based on the article Magn Reson Med 2003; Mar, 49 (3): 403-8.

Recently a novel Water Soluble Iron Oxide (WSIO) nanoparticle has been designed and synthesized [67]. This nanoparticle was conjugated to Herceptin, and for the first time *in vivo* cancer targeting and imaging of HER2/neu were performed using this magnetic nanocrystal probe (Fig. **12**) [67]. The NIH3T6.7 cell line, which overexpresses the HER2/neu cancer marker, was implanted in a set of nude mice. After tail vein injection of the WSIO-Herceptin conjugate, **5**, MR imaging of the mice was performed at different time points (*e.g.*, preinjection, immediately postinjection (~5 minute), 1h, 2h, and 4h). The $T_2$ value drop was observed with time; ~10% drop within 5 min, ~15% drop in 2h, and ~20% drop in 4h, and a color change from green to blue was also observed in the color mapped MRI signal [67]. On the other hand, in the control experiment with WSIO-irrelevant antibody (anti-HMEN-B3) conjugate, no change in the $T_2$ values was observed at the tumor site, and also no change in the color-mapped MRI signal. These results are consistent with *in vitro* studies [67]. Moreover, this WSIO-Herceptin probe conjugate was used to monitor the evolution of the events of active targeting *in vivo* at higher magnetic field (9.4 T) [67]. This novel WSIO nanocrystal and its use in MR imaging can be applied for *in vitro* and *in vivo* diagnosis of other types of cancers as well.

Although the specificity of antibody-IO nanoparticle conjugate as target-specific tumor imaging probes has been demonstrated, such nanoparticle also have several limitations. For example, i) the large size of antibody may affect the efficacy of conjugation, ii) various residues of the antibody can participate in the conjugation giving multiple products, iii) conjugation of the antibody to the nanoparticle may affect the binding properties of the antibody, iv) the large size of the antibody-nanoparticle conjugate may reduce its permeation through the vasculature into tumor cells, v) the specificity of tumor-targeted nanoparticles can decrease due to the interaction of antibody with Fc (fragment, crystallizable) receptors on normal tissues. Thus, peptides and low molecular weight single chain antibodies can be better targeting ligands for constructing target-specific IO nanoparticles.

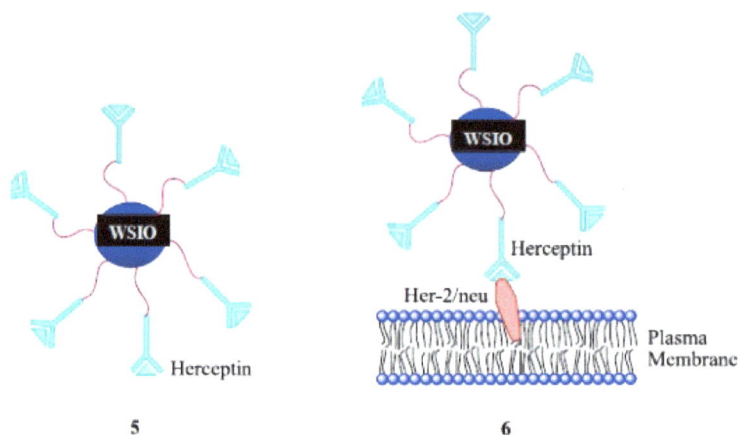

Herceptin

Her-2/neu

Herceptin

Plasma
Membrane

5

6

**Figure 12:** WSIO nanocrystal-Herceptin conjugate. This figure is drawn based on the article J Am Chem Soc 2005; Sep 7, 127 (35): 12387-91.

### 3.4 Use of Iron Oxide Nanoparticles and Quantum Dots in Targeting of EGFR

Iron oxide nanoparticles and quantum dots have been used in EGFR-targeted MR imaging and optical imaging respectively. A single-chain anti-EGFR antibody ScFvEGFR has been conjugated to surface-functionalized QD and iron oxide (IO) to produce the nanoparticles ScFvEGFR Ab-QD (Fig. **13**, compound **7**) and ScFvEGFR Ab-IO (Fig. **14**, compound **8**) respectively, for *in vivo* targeting and imaging of an orthotopic pancreatic cancer xenograft model [34].

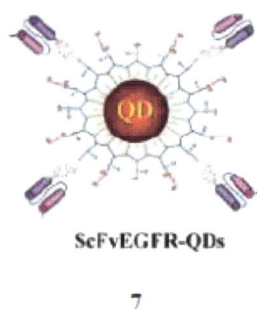

ScFvEGFR-QDs

7

**Figure 13:** QD-ScFvEGFR antibody-conjugated nanoparticle. Reproduced with permission from Yang L, Mao H, Wang YA, Cao Z, Peng X, Wang X, *et al.* Single chain epidermal growth factor receptor antibody conjugated nanoparticles for *in vivo* tumor targeting and imaging. Small 2009; 5 (2): 235-43.

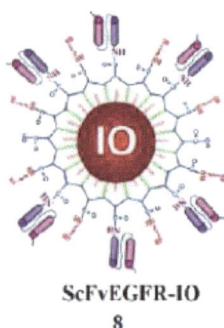

ScFvEGFR-IO

8

**Figure 14:** IO-ScFvEGFR antibody-conjugated nanoparticle. Reproduced with permission from Yang L, Mao H, Wang YA, Cao Z, Peng X, Wang X, *et al.* Single chain epidermal growth factor receptor antibody conjugated nanoparticles for *in vivo* tumor targeting and imaging. Small 2009; 5 (2): 235-43.

EGFR is overexpressed in various cancers, such as pancreatic cancer (MIA PaCa-2 cells), breast cancer (MDA-MB-231 cells, 4T1 cells), and head and neck cancer (Tu212 cells). In an effort to determine the specificity of the nanoparticle ScFvEGFR Ab-QD **7**, a binding study was performed in MIA PaCa-2, MDA-MB-231, and 4T1 cells [34]. When incubated at 4°C, ScFvEGFR Ab-QD **7** was selectively bound to the cell surface while non-targeted QDs did not bind to the cells. Uptake of ScFvEGFR Ab-QD **7** into MDA-MB-231 cells was confirmed by confocal microscopy. In a separate experiment, ScFvEGFR Ab-QD **7** was injected into mice bearing orthotopic pancreatic cancer xenografts from the EGFR-positive human pancreatic cancer cell line, MIA PaCa-2 [34]. Localization of ScFvEGFR Ab-QD **7** nanoparticle in the cytoplasm of the pancreatic tumor mass occurred, probably *via* receptor-mediated internalization.

EGFR-targeted *in vivo* MRI using the ScFvEGFR Ab-IO nanoparticle (compound **8**) was also studied in an orthotopic pancreatic cancer xenograft model [34]. ScFvEGFR Ab-IO **8** was administered through the tail vein, and MR imaging was performed at various pre and post-contrast time points. $T_2$-weighted fast spin echo imaging results demonstrated that the ScFvEGFR Ab-IO **8** accumulated selectively inside the pancreatic tumor, evidenced by the marked MRI contrast decrease in the pancreatic tumor area (Fig. **15**).

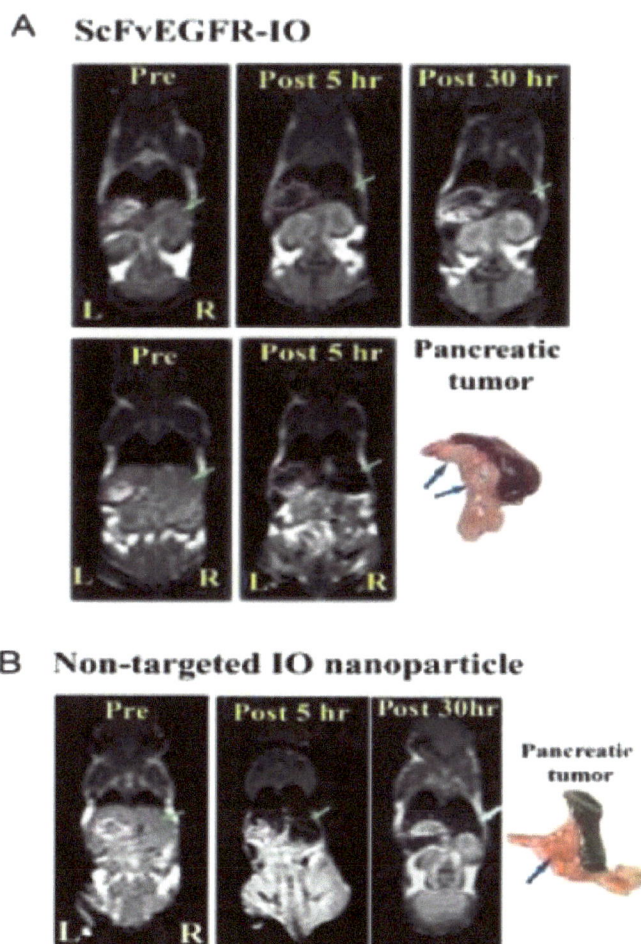

**Figure 15:** Examination of target specificity of ScFvEGFR-IO nanoparticles by MRI using an orthotopic human pancreatic xenograft model. A) MRI of a tumor-bearing mouse. ScFvEGFR-IO nanoparticles (8 nmol kg$^{-1}$ body weight) were injected into the mouse through the tail vein. Pre- and post-contrast MRI at 5 and 30 h were collected. Upper and lower panels showed the MRI from different sectional levels of the same mouse. The areas of the pancreatic tumor were marked as a dash-lined circle (pink). The pancreatic tumor area showed a bright signal before receiving the nanoparticle. After injection of the targeted IO nanoparticles, a marked MRI contrast decrease was detected in the tumor (darker), which delineated the area of the tumor lesion. MRI contrast change was also found in the liver (green arrow) and spleen. These MRIs are representative results of five mice that received ScFvEGFR-IO nanoparticles.

Lower right is the picture of tumor and spleen tissues, showing sizes and locations of two intra-pancreatic tumor lesions (arrows) that correspond with the tumor images of MRI. B) MRI of a mouse that received non-targeted IO nanoparticles. The tumor area (pink dash-lined circle) did not show MRI signal decrease at 5 and 30 h after the injection of nanoparticles. The areas representing the liver and spleen had marked signal decrease due to the $T_2$ contrast. Shown are representative MRIs of three mice that received control IO nanoparticles. Reproduced with permission from Yang L, Mao H, Wang YA, Cao Z, Peng X, Wang X, *et al.* Single chain epidermal growth factor receptor antibody conjugated nanoparticles for *in vivo* tumor targeting and imaging. Small 2009; 5 (2): 235-43.

### 3.5 Use of Iron Oxide Nanoparticles in Targeting of uPAR

Urokinase Plasminogen Activator (uPA) is a serine protease that interacts with its cellular receptor (urokinase Plasminogen Activator Receptor, uPAR) and regulates matrix degradation, cell invasion, and angiogenesis. uPAR is a cellular receptor highly overexpressed in various cancers such as pancreatic cancer, breast cancer and lung cancer. uPAR is highly overexpressed in more than 86% of pancreatic cancer tissue, but is absent in pancreatic tissue obtained from healthy subjects. A recombinant peptide containing the amino-terminal fragment (ATF) of uPA can bind uPAR with high affinity. In fact, ATF can compete with uPA for binding uPAR at the surface of both tumor and endothelial cells, resulting in inhibition of tumor growth and angiogenesis. uPAR is overexpressed in pancreatic cancer cell lines such as MIA PaCa-2.

The nanoparticle ATF-IO has been synthesized by conjugating the ATF peptide with iron oxide (IO) nanoparticles [68]. An orthotopic human pancreatic cancer xenograft model was established by injecting 5 x $10^6$ MIA PaCa-2 cells into the pancreas of female nude mice. Systemic delivery of ATF-IO nanoparticles led to their selective accumulation in the tumors of these mice, which was confirmed by near infrared optical imaging and MRI [68].

uPAR is also overexpressed in breast cancer cells such as the 4T1 cell line. *In vivo* MR imaging of breast cancer has also been performed successfully using ATF-IO nanoparticles [68].

Our research group has purified the Amino-Terminal Fragment (ATF, Molecular weight: 14 kD) of uPA and conjugated ATF to IO nanoparticles. We observed that the systemic delivery of ATF-IO nanoparticles into mice bearing mammary tumors led to the accumulation of nanoparticles selectively in tumors and generated strong MRI contrast, while the nonspecific uptake by the liver and spleen decreased significantly in comparison to that in mice injected with non-targeted IO nanoparticles (Fig. **16**) [69]. By conjugating a near-IR dye, Cy5.5 to the ATF-peptide, we have accomplished *in vivo* tumor imaging in two modalities, optical imaging as well as MRI. These results demonstrate the successful application of iron oxide nanoparticles in MR imaging of pancreatic cancer and breast cancer based on targeting of uPAR.

### 3.6 Other Targeting Approaches Using Iron Oxide Nanoparticles

Other peptides have also been used as targeting ligands in tumor targeted IO nanoparticles to image tumor, metastasis and tumor vessels *in vivo* [70-72].

Most of the currently used target molecules, such as EGFR, Her-2/neu and PSMA (Prostate Specific Membrane Antigen) are overexpressed only in subpopulations of tumor tissues or specific tumor types, limiting their application in imaging and therapy.

In order to deliver sufficient amount of IO nanoparticles in both the tumor vessels and tumor stroma for achieving high sensitivity, Simberg *et al.* conjugated the tumor-homing peptide, Cys-Arg-Glu-Lys-Ala (CREKA) to IO nanoparticles and showed that these nanoparticles could accumulate in both tumor vessels and stroma, induce intravascular clotting and generate new binding site for more nanoparticles in the tumor, further amplifying the target [73]. These novel targeted-IO nanoparticles have the potential to increase the sensitivity and specificity of tumor detection by optical imaging techniques [73].

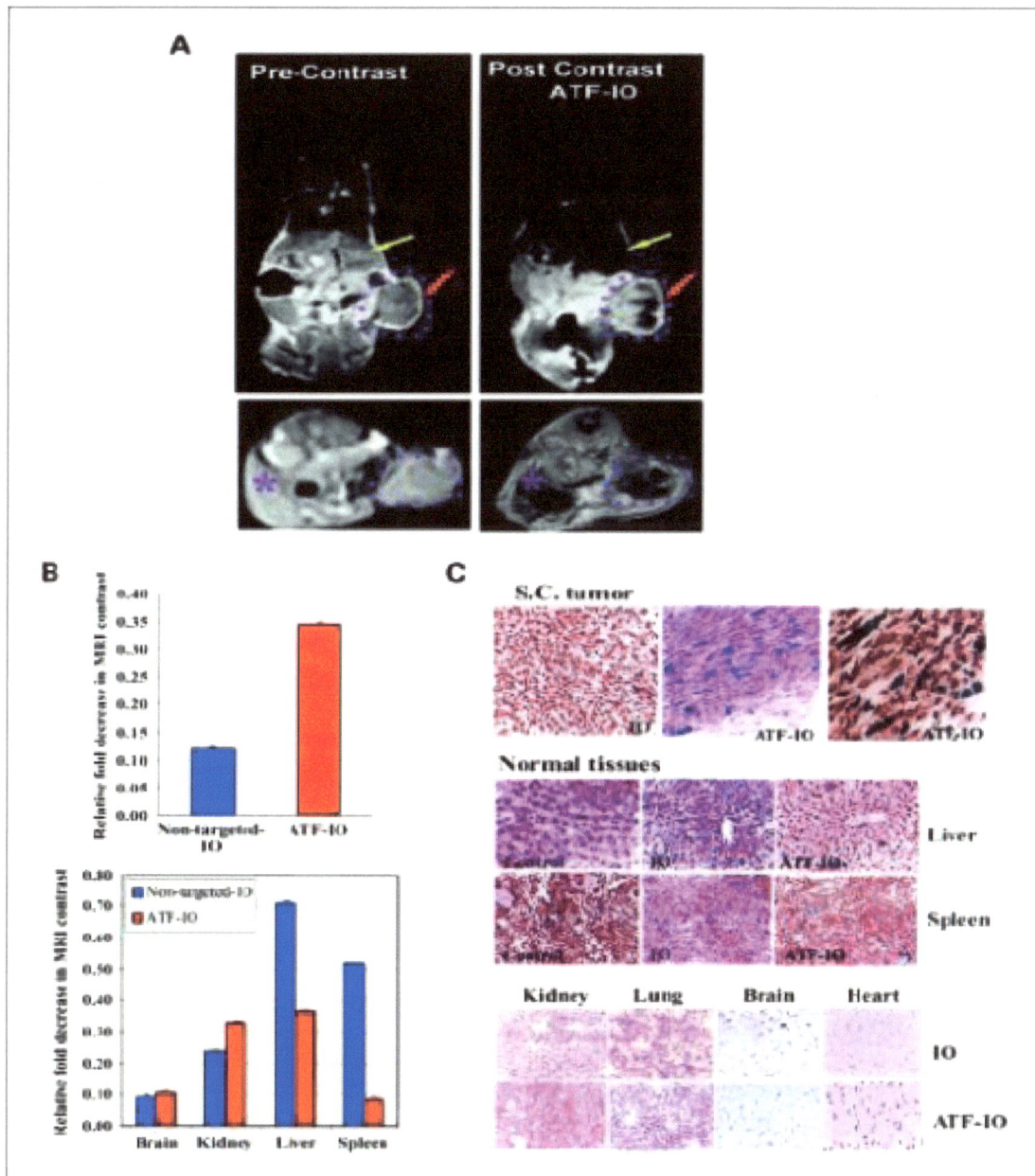

**Figure 16:** *In vivo* magnetic resonance imaging of $4T_1$ mammary tumor using ATF-IO nanoparticles. (A) magnetic resonance imaging of s.c. mammary tumor. A marked magnetic resonance imaging signal drop with $T_2$ weighted contrast was observed in s.c. tumor areas (pink dashed lined) 6 h after tail vein injection of ATF-IO nanoparticles. Heterogeneous signal changes suggest that intratumoral distribution of ATF-IO nanoparticles was not uniform in the s.c. tumor (red arrow). $T_2$ contrast change was also found in the liver (yellow arrows and pink asterisk). Selected magnetic resonance image is a representative image of seven mice that received ATF-IO nanoparticles. (B) Magnetic resonance image of the mouse that received nontargeted iron oxide nanoparticles, organ-specific profiling of magnetic resonance imaging signal change. Signal changes in the mice that received nontargeted iron oxide or ATF-IO nanoparticles for 6 h were measured in the regions of tumor or various normal tissues. Relative intensity was calculated using the intensity in the leg muscle as a reference. Fold decreases in the intensity of the magnetic resonance image were compared between pre– and post–ATF-IO nanoparticle injections and plotted in the figure. Bar plot, mean values of three regions. (C) Prussian blue staining of B iron oxide nanoparticles in tumor and normal tissues 48 h after injection of the nanoparticles. Red, background staining with nuclear fast red. Reproduced with permission from Yang L, Peng XH, Wang YA, Wang X, Cao Z, Ni C, *et al.* Receptor-targeted nanoparticles for *in vivo* imaging of breast cancer. Clin Cancer Res 2009; 15 (14): 4722-32.

Folate receptors are overexpressed in approximately 40% of human cancers. Folic acid (FA) has low a molecular weight and high binding affinity for folate receptor ($K_d = 10^{-10}$ M). It is low-priced, non-immunogenic, and amenable to easy conjugation. Hence, FA has been extensively used as a targeting ligand in targeted tumor imaging and therapy [74]. Folate receptor-positive human cervical carcinoma HeLa cells could take up FA-IO nanoparticles to 12-fold greater levels than that of non-targeted IO nanoparticles *in vitro* [75]. FA-targeted IO nanoparticles were also able to target human nasopharyngeal epidermoid carcinoma cells both *in vitro* and *in vivo* [76].

### 3.7 Advantages and Disadvantages of Iron Oxide Nanoparticles

Although MRI can provide high spatial resolution without any limitation in penetration depth, its sensitivity is low ($10^{-3}$-$10^{-5}$ M). In contrast, conventional optical imaging has a high sensitivity ($10^{-9}$-$10^{-12}$ M), but has low depth penetration (<1 cm) and low resolution.

Despite much progress in the development of IO nanoparticle as MRI contrast agents, one of the major challenges is to develop a surface coating material that can stabilize the nanoparticles and provide functional groups for conjugation of targeting ligands. When traditional ligands such as dextran were used to coat the IO nanoparticles to stabilize the IO, the ligand-particle interaction usually became weak, resulting in separation of the ligands from the surface of the nanoparticles, and further causing aggregation and precipitation of nanoparticles under physiological conditions or simply during storage. This weak interaction between ligand and nanoparticles may not be able to withstand various reaction conditions.

Recently, our group has developed a new class of magnetic IO nanoparticles that have uniform sizes ranging from 5-30 nm and can be further functionalized by surface coating with amphiphilic triblock polymers, which provide functional groups for conjugating tumor-targeting biomolecules such as peptides or antibodies. The carboxylic acid groups on the nanoparticle surface make the particles hydrophilic, stable and amenable to bioconjugation. Our previous data showed that this coating withstood the *in vivo* conditions and preserved the antibody or peptide conjugated with the IO nanoparticles for effective targeting of the tumor [27, 69].

Future studies will focus on the development of tumor-targeted IO nanoparticles having high sensitivity and specificity that can be used in the early stage detection of tumor, monitoring of tumor metastasis and its response to therapy.

### 4. MULTIMODALITY IMAGING

Multimodality imaging has become extremely popular in the field of cancer diagnosis, since it allows the evaluation of the same target with multiple modalities, such as Positron Emission Tomography (PET), MRI, Computed Tomography (CT), and near infrared spectroscopy optical imaging. Hence, the advantages of multiple imaging modalities can add up to improve the overall diagnostic accuracy. Unlike single modality imaging, multimodality imaging can provide more precise information regarding a disease, such as its location, extent, metabolic activity, blood flow, function of target tissue etc, resulting in better characterization of disease processes.

### 4.1 PET/CT, Fluorescence Molecular Tomography, and MR Imaging

Advancement in nanotechnology has had profound impact on the field of molecular imaging. In particular, substantial progress has been made in the development of novel nanoparticles suitable for use in multimodal imaging.

Recently an [18]F-modified trimodal nanoparticle which consists of an iron oxide core coated with aminated dextran has been synthesized (Fig. **17**, compound **9**) [77]. The near IR fluorochrome Vivotag-680 and the radionuclide [18]F have been conjugated to the amine functionality on the dextran [77]. Azide functionalities on the surface of the aminated Cross-Linked Dextran Iron Oxide (CLIO) nanoparticle can be used to conjugate various targeting ligands for targeted cancer imaging [77]. These trimodal bifunctional nanoparticles can be detected with PET, MRI, fluorescence molecular tomography, and are suitable for imaging as well as therapy.

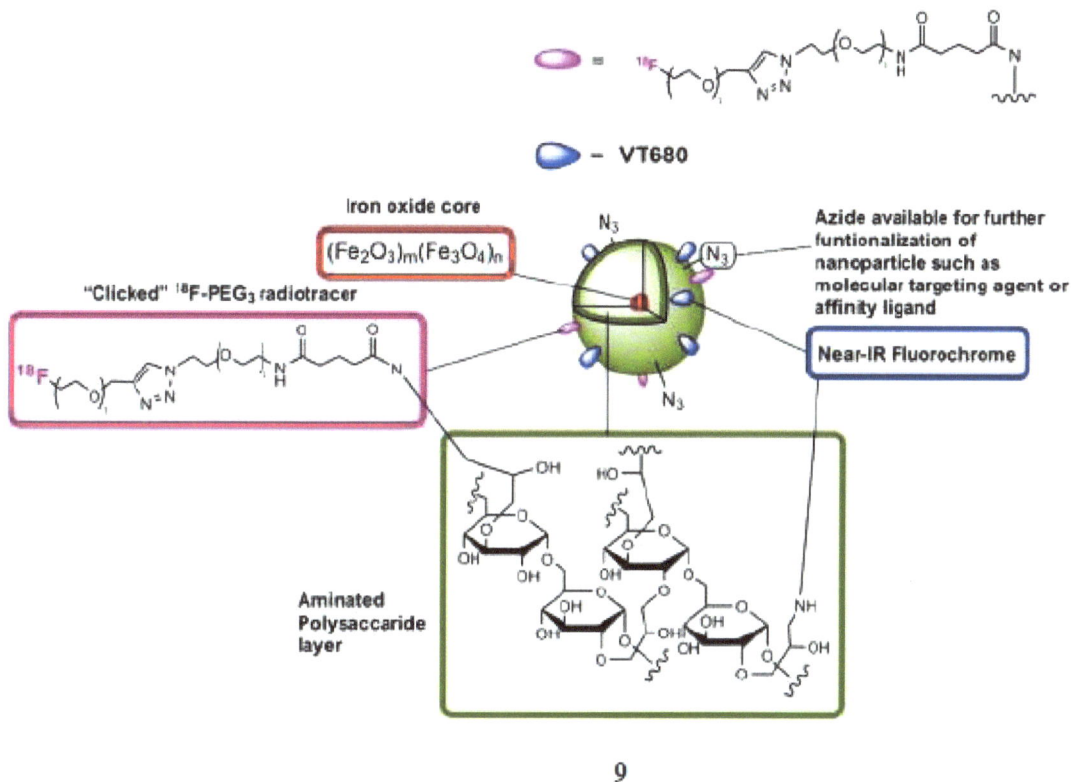

**9**

**Figure 17:** $^{18}$F-CLIO-Vivotag-680 nanoparticles for multimodal imaging. Reproduced with permission from Devaraj NK, Keliher EJ, Thurber GM, Nahrendorf M, Weissleder R. $^{18}$F labeled nanoparticles for *in vivo* PET-CT imaging. Bioconjug Chem 2009; Feb, 20 (2): 397-401.

*In vivo* dynamic PET/CT imaging of BALB/C mouse after injection of the $^{18}$F-CLIO nanoparticles **9**, demonstrated a very high signal to noise ratio [77]. These nanoparticles have a vascular half-life of 5.8 hours. Ultimately, these nanoparticles are internalized into the macrophages of liver, spleen, and phagocytic cells. They are biodegradable, and slowly decompose into elemental components. The iron is metabolized and reused in the synthesis of heme. The dextran coating is slowly excreted.

Conjugates based on this trimodal nanoparticle and a targeting ligand, combined with an optimal pharmacokinetic profile, may be attractive candidates for clinical molecular imaging.

## 4.2 Optical and Magnetic Resonance (MR) Imaging Based on Inverse Targeting Approach

An inverse targeting approach termed as a "normal tissue-targeted nanoparticle/$T_2$ reduction strategy" has been used to image pancreatic ductal adenocarcinoma (PDAC) [78]. The 14-amino acid peptide bombesin (BN) is a tumor marker for pancreatic cancer, small cell lung carcinoma and neuroblastoma. A structurally unique nanoparticle, BN-CLIO-Cy5.5, composed of BN, Cross-Linked Iron Oxide (CLIO), and the fluorescent dye Cy5.5 has been synthesized [78]. Quantitative tissue fluorescence experiments demonstrated that the BN-CLIO-Cy5.5 nanoparticles bind selectively to the BN receptors present on normal acinar cells of the pancreas, but not to the Pancreatic Ductal Adenocarcinoma (PDAC) tissue. MIA-PaCa2 cells were implanted in the pancreatic tail of athymic female nude mice to generate pancreatic tumor. BN-CLIO-Cy5.5 was administered through tail vein injection, and mice were imaged after 24 h in a

4 T Bruker system [78]. The BN-CLIO-Cy5.5 nanoparticle specifically decreased the $T_2$ of normal pancreas, resulting in an enhanced ability to visualize the tumor by MR imaging [78]. However, the observed heterogeneous changes in the $T_2$ of the normal pancreas were probably due to a heterogeneous vasculature, limiting the nanoparticle delivery to acinar cells.

## 5. THERAGNOSTICS: DIAGNOSTICS AND THERAPY

Theragnostics is a recently approach to cancer treatment based on the fusion between diagnostics and therapeutics [12-16]. In this treatment strategy, the results of diagnostic tests are interlinked with targeted therapy. Advances in genomics, proteomics, and bioinformatics have generated new information on genes, proteins and biomarkers that have enabled the development of new diagnostic assays. In theragnostics, the results of diagnostic tests are used to design and select the most appropriate therapy, in which an anti-cancer drug is delivered selectively to the targeted cancer site with the help of targeting ligands such as antibodies, small molecules, peptides, peptidomimetics and aptamers. Hence, diagnostics and therapy work hand in hand. Moreover, theragnostics aims to monitor the treatment responses, increasing drug efficacy and safety. Immense effort is being made to develop agents that can potentially be used in targeted cancer imaging as well as therapy.

### 5.1 Theragnostic Superparamagnetic Iron Oxide Nanoparticles for Targeting of Folate Receptors

A unique water-based methodology has been employed to synthesize biocompatible, multimodal, multifunctional theragnostic superparamagnetic iron oxide nanoparticles. In these nanoparticles, the superparamagnetic iron oxide core is coated with polyacrylic acid. An anticancer drug Paclitaxel and a lipophilic near infrared (NIR) dye (dialkyl carbocyanine fluorescent dye) have been co-encapsulated in the hydrophobic matrix of the polyacrylic acid [79]. The carboxylic acid group of polyacrylic acid was reacted with propargyl amine to obtain acetylene functionality on the nanoparticle surface. This acetylene functionality was conjugated with azide-functionalized folic acid to obtain the multimodal, multifunctional nanoparticle-folic acid conjugates (Fig. **18**, compound **10** and Fig. **19**, compound **11**) [79]. These nanoparticles have demonstrated properties suitable for optical and MR imaging, and for targeted therapy of cancer. Compound **10** is appropriate for targeted imaging of cancer, and compound **11** is appropriate for targeted optical and MR imaging as well as therapy [79]. The average diameter of these nanoparticles is about 90 nm.

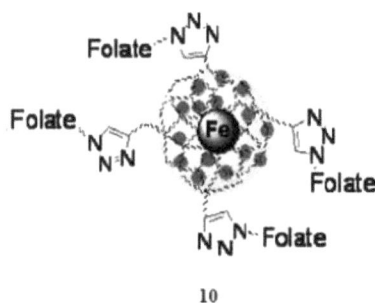

10

**Figure 18:** Schematic representation of theragnostic iron oxide nanoparticles. Reproduced with permission from Santra S, Kaittanis C, Grimm J, Perez JM. Drug/dye-loaded, multifunctional iron oxide nanoparticles for combined targeted cancer therapy and dual optical/magnetic resonance imaging. Small 2009; Aug 17, 5 (16): 1862-8.

In order to examine the *in vitro* differential cytotoxicity of compounds **10** and **11**, MTT assay was performed using lung carcinoma (A549 cell) and cardiomyocyte (H9c2) cell lines [79]. Folate receptor is overexpressed in the A549 cell line but not in the H9c2 cell line. The use of compound **11** led to an 80% reduction in the viability of the A549 lung cancer cell line. On the contrary, no significant reduction in cell viability was observed when H9c2 cardiomyocyte cells were incubated with compound **11**. Compound **10** showed much less cytotoxicity (less than 3% in comparison to the control) towards both the cell lines A549 and H9c2.

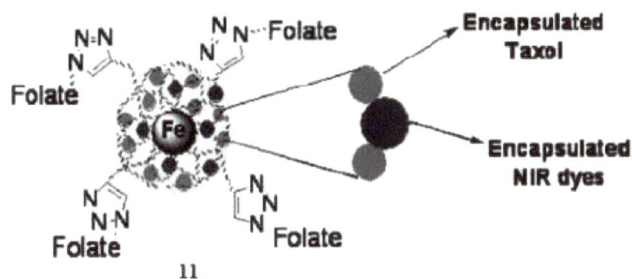

**Figure 19:** Schematic representation of theragnostic iron oxide nanoparticles. Reproduced with permission from Santra S, Kaittanis C, Grimm J, Perez JM. Drug/dye-loaded, multifunctional iron oxide nanoparticles for combined targeted cancer therapy and dual optical/magnetic resonance imaging. Small 2009; Aug 17, 5 (16): 1862-8.

In a separate experiment, the cellular uptake of folate functionalized nanoparticle **10** by A549 lung cancer cells and its internalization was confirmed using confocal laser scanning microscopy [79]. The cytoplasm of A549 cells displayed strong fluorescence.

After the intracellular uptake, the release of Paclitaxel from nanoparticles occured probably *via* the esterase-mediated degradation or in acidified lysosomes.

These biocompatible, water soluble, multimodal and multifunctional theragnostic iron oxide nanoparticles have potential application in targeted imaging and targeted therapy of cancer.

### 5.2 Manganese-Labeled Theragnostic Agent for MR Imaging and Delivery of Anticancer Drugs

Novel manganese (III)-labeled nanobialys nanoparticles have been synthesized recently by molecular self-assembly of amphiphilic branched polyethylenimine (Fig. **20**, compound **12**) [80]. These "bialy" shape nanoparticles have a toroidal shape, high stability, tunable and uniform particle size, and present the stable Mn(III)-porphyrin complexes directly to the surrounding water [80]. The average diameter of these nanoparticles is about 190 nm. These Mn (III)-labeled nanobialys nanoparticles are appropriate for theragnostics applications, for both MR imaging as well as targeted delivery of anticancer drugs [80].

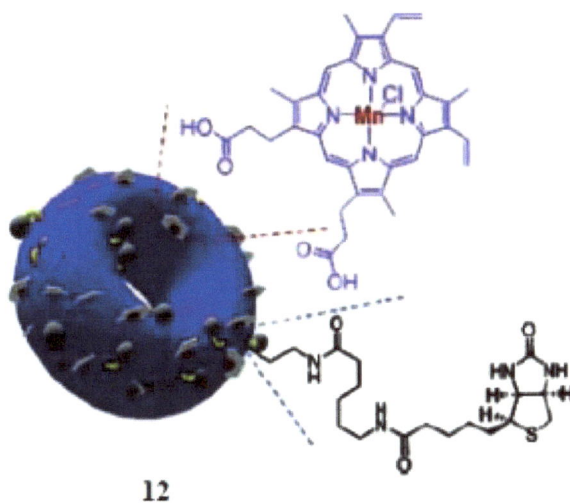

**Figure 20:** Schematic representation of manganese-labeled nanobialys. Reproduced with permission from Pan D, Caruthers SD, Hu G, Senpan A, Scott MJ, Gaffney PJ, *et al.* Ligand-directed nanobialys as theranostic agent for drug delivery and manganese-based magnetic resonance imaging of vascular targets. J Am Chem Soc 2008; Jul 23; 130(29): 9186-7.

The r$_1$ and r$_2$ relaxivities of Mn (III)-labled nanobialys were 3.7±1.1 (s.mmol [Mn])$^{-1}$ and 5.2±1.1 (s.mmol [Mn])$^{-1}$ respectively and the particulate relaxivities were 612307±7213 (s.mmol [nanobialy])$^{-1}$ and 866989±10704 (s.mmol [nanobialy])$^{-1}$ respectively [80]. MR molecular imaging of fibrin was performed *in vitro* at a field strength of 3T. T$_1$ weighted gradient echo images of clot samples displayed marked contrast enhancement of the fibrin-targeted Mn (III) nanobialys, whereas no contrast enhancement was observed in the case of non-targeted Mn-free nanobialys [80].

Anticancer drugs doxorubicin and camptothecin have been encapsulated in the nanobialys. Studies showed that 12±0.6% of doxorubicin and 20±3.5% of camptothecin were released over 3 days [80].

The Mn-based nanobialys demonstrated the potential for site-specific T$_1$-weighted MR molecular imaging as well as local delivery of cancer chemotherapeutics. *In vivo* studies and subsequent optimization will definitely help to develop a theragnostic agent ideal for clinical applications.

### 5.3 Gadolinium Metal-Organic Framework Based Theragnostic Agents for Targeting of α$_5$β$_3$ Integrins

The surface of Gadolinium Metal-Organic Framework (Gd-MOF) nanoparticles has been modified by covalent conjugation of well-defined copolymers of poly-(N-isopropylacrylamide)-co-poly(N-acryloxysuccinimide)-co-poly(fluorescein O-methacryl-ate) PNIPAM-co-PNAOS-co-PFMA (Fig. **21**) [81]. This Gd-MOF-RAFT copolymer was conjugated to the targeting ligand glycine-arginine-glycine-aspartate-serine-NH$_2$ (GRGDS-NH$_2$) peptide and the antineoplastic drug methotrexate (MTX) to give Gd-MOF-RAFT copolymer PNIPAM-co-PNAOS-co-PFMA-GRGDS-NH$_2$ conjugate **13** and Gd-MOF-RAFT copolymer PNIPAM-co-PNAOS-co-PFMA-MTX conjugate **14**, respectively (Fig. **21**) [81]. Gd is extensively used in T$_1$ weighted MR imaging, whereas fluorescein O-methacrylate is used in optical imaging. The GRGDS-NH$_2$ peptide can specifically bind to the α$_5$β$_3$ integrins over expressed on tumor cells, and induce the cellular uptake of the nanoparticle by receptor mediated endocytosis. Compound **13** is designed for targeted optical imaging and MR imaging, whereas compound **14** is appropriate for targeted therapy as well as dual (MR and optical) imaging [81]. Compound **15** composed of the polymer, imaging agents (gadolinium and fluorescein), targeting peptide ligand and the drug methotrexate (Fig. **21**), and is thus designed for targeted dual (MR and optical) imaging as well as targeted therapy [81]. These agents offer the high sensitivity of fluorescence imaging as well as the high degree of spatial resolution of MR imaging.

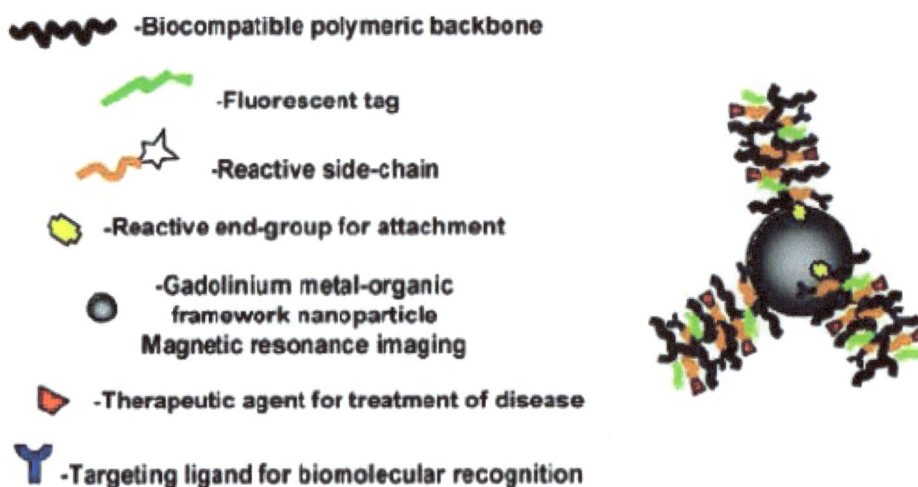

**Figure 21:** Polymer-modified Gd metal-organic framework nanoparticles as a nanomedicine construct for targeted imaging and treatment of cancer. Reproduced with permission from Rowe MD, Thamm DH, Kraft SL, Boyes SG. Polymer-modified gadolinium metal-organic framework nanoparticles used as multifunctional nanomedicines for the targeted imaging and treatment of cancer. Biomacromolecules 2009; Apr 13, 10 (4): 983-93.

The relaxivity of Gd-MOF-PNIPAM-*co*-PNAS-*co*-PFMA-GRGDS-NH$_2$ nanoparticles **13** ($r_1$ = 33.43 mM$^{-1}$s$^{-1}$) is more than three times in comparison to the unmodified Gd-MOF nanoparticles ($r_1$ = 9.86 mM$^{-1}$s$^{-1}$) (Table **1**) due to the increased water retention by the hydrophilic raft copolymer matrix attached to the surface of the Gd-MOF nanoparticles [81]. The relaxivity of RAFT copolymer-modified Gd-MOF nanoparticles **13** ($r_1$ = 33.43 mM$^{-1}$s$^{-1}$) is also much higher than that of the clinically-used Gd contrast agents Magnevist ($r_1$ = 13.44 mM$^{-1}$s$^{-1}$) and Multihance ($r_1$ = 19.45 mM$^{-1}$s$^{-1}$). At a concentration, in the range of [Gd$^{3+}$] = 0.013-0.16 mmol/L, the relaxivity of MTX-GRGDS-NH$_2$-containing Gd-MOF–polymer-MTX nanoparticle **15** is 14.45 mM$^{-1}$s$^{-1}$, whereas at a concentration, in the range of [Gd$^{3+}$] = 0.046-0.51 mmol/L, the relaxivity of Magnevist is 13.44 mM$^{-1}$s$^{-1}$. Relaxivity is nearly proportional to the concentrations of Gd$^{3+}$. This result demonstrates that at a similar concentration of Gd$^{3+}$, the polymer-modified Gd-MOF nanoparticles will produce comparable or even higher relaxivity than that of clinically-used Magnevist or Multihance.

**Table 1:** Experimental relaxivity data of MR imaging agents [81]

| Contrast agent | $r_1$ (mM$^{-1}$s$^{-1}$) | $r_2$ (mM$^{-1}$s$^{-1}$) |
|---|---|---|
| Magnevist | 13.44 | 21.40 |
| Multihance | 19.45 | 30.44 |
| Gd-MOF nanoparticles | 9.86 | 17.94 |
| Gd-MOF-PNIPAM-*co*-PNAS-*co*-PFMA-GRGDS-NH$_2$ (**13**) | 33.43 | 47.25 |
| Gd-MOF-PNIPAM-*co*-PNAS-*co*-PFMA-MTX (**14**) | 38.52 | 53.92 |
| Gd-MOF-PNIPAM-*co*-PNAS-*co*-PFMA-MTX+GRGDS-NH$_2$ (**15**) | 14.45 | 25.29 |

Gd-MOF-RAFT copolymer-GRGDS-NH$_2$-MTX conjugate **15** was incubated with FITZ-HSA, an $\alpha_v\beta_3$-expressing canine endothelial sarcoma cell line, followed by fluorescence imaging at various time points [81]. The GRGDS-NH$_2$-MTX-tailored RAFT polymer-modified Gd-MOF nanoparticles were internalized after an incubation period of 24 h. When Gd-MOF-PNIPAM-*co*-PNAS-*co*-PFMA-MTX conjugate **14** was incubated with FITZ-HSA, no internalization was observed even after 24 h due to the absence of GRGDS-NH$_2$ in the conjugate **14**. These results evidently demonstrated the preferential uptake of the targeting ligand containing polymer modified Gd-MOF nanoparticles **15** was due to the active targeting of $\alpha_v\beta_3$-integrins by the GRGDS-NH$_2$ peptide.

Cell growth inhibition properties of polymer-modified Gd-MOF nanoparticles were investigated using FITZ-HSA tumor cells [81]. MTX-tailored polymer-modified Gd-MOF nanoparticles **14** showed a dose-dependent inhibition of growth of the FITZ-HSA tumor cells that was comparable to that of the free MTX drug, on the basis of equal concentration of MTX. This result demonstrates that even after the attachment of MTX to the polymer and subsequent modification of the Gd-MOF nanoparticles with MTX copolymer the overall efficacy of the drug is still retained [81].

Since these theragnostics nanoparticles are ideal for fluorescence and MR imaging as well as for targeted cancer therapy, they are potential candidates for clinical applications.

### 5.4 Theragnostic Nanoparticles for MR Mapping of Tumor Angiogenesis

The $\alpha_5\beta_1$ ($\alpha_v\beta_3$)-targeted theragnostic nanoparticles (rhodamine-labeled and perfluorocarbon nanoparticles) have recently been used for three dimensional (3D) MR mapping of angiogenesis in the MDA-MB-435 xenograft mouse model [82]. The RGD peptide and peptidomimetic were employed in ligand-targeted MR molecular imaging to target the $\alpha_5\beta_1$ ($\alpha_v\beta_3$)-integrins implicated in angiogenesis of MDA-MB-435 tumors. 3D mapping of tumor neovascularity demonstrated that 90% of the new vessel signal was distributed peripherally, and they were asymmetric and generally sparse, with no neovessel signal increment over large portions of the tumor surface [82]. Immunohistochemical analysis demonstrated that the targeted rhodamine-labeled nanoparticles were bound and co-localized with FITC-lectin only in the peripheral microvasculature, but not in the tumor core. It was noticed that the $\alpha_5\beta_1$-targeted fumagillin nanoparticles decreased the sparse neovascularity more than that of the $\alpha_v\beta_3$-targeted nanoparticle counterpart relative to control [82]. The MDA-MB-435 tumor model had a sparse neovasculature and there was no growth response to antiangiogenic therapy even after reduction in peripheral neovessels. These results imply that theragnostic nanoparticle-based MR molecular imaging of angiogenesis can identify tumors having low levels of neovasculature which may have poor response to antiogenic therapy [82].

## 6. STRATEGIES TO REDUCE THE NONSPECIFIC UPTAKE OF NANOPARTICLES

Despite the advantages of using targeted nanoparticles (QDs, gold nanoparticles, IO nanoparticles) in non-invasive imaging *in vivo*, the specificity and sensitivity are still affected mainly by the non-specific uptake of nanoparticles by liver (Küpffer cells) and spleen, since the body always treat the nanoparticles as foreign materials [83, 84]. Macrophages can cause internalization of a wide variety of nanoparticles including QDs, IO and gold nanoparticles. There are many factors involved in the regulation of the non-specific uptake of nanoparticles by macrophage cells [71, 73, 85-92].

Macrophage function is regulated by cytokines such as interferon-gamma (IFN-$\gamma$), interleukin-4 (IL-4), and serum proteins. Lovastatin, the inhibitor of 3-hydroxy-3-methylglutaryl coenzyme A (HMG-CoA) reductase, is believed to downregulate the macrophage scavenger receptors. Rogers and Basu have shown that the dextran-coated superparamagnetic iron oxide (SPIO) nanoparticle uptake by activated macrophages is enhanced by the cytokines IL-4 and IFN-$\gamma$ [85]. On the contrary, lovastatin has significantly down-regulated the macrophage-mediated endocytosis of dextran-coated SPIO nanoparticles [85]. In addition to being an MRI probe, dextran-coated SPIO nanoparticles can provide insight into the changes in macrophage function associated with anti-inflammatory therapy.

Decoy particles such as Ni-liposome and liposomal clodronate can deplete the RES macrophages from the circulation. Elimination of macrophages by injecting decoy particles can reduce the non-specific uptake of nanoparticles, resulting in the prolonged half-life of nanoparticles. A recent study has demonstrated that intravenously injected Ni-liposomes prolonged the half-life of SPIO nanoparticles and CREKA-SPIO nanoparticles in the blood by a factor of ~5 [73]. The Ni-liposome pretreatment performed before the injection of CREKA-SPIO, greatly increased the tumor homing of the nanoparticles, which were mainly localized in the tumor blood vessels [73].

SPIO nanoparticles (SPION) have already demonstrated their immense potential in biomedical applications such as MRI, drug delivery, and hyperthermia therapy. However, their application has become limited due to biofouling of the particles in the blood plasma, and aggregate formation that is sequestered by the cells of the RES. Moreover, aggregation of SPION affects its intrinsic superparamagnetic properties. It has become necessary to modify the surface of the SPION to minimize the biofouling and particle aggregation under physiological conditions.

Hence, the poly(TMSMA-*r*-PEGMA)-coated SPION (Fig. **22**, compound **16**) was synthesized and evaluated as an *in vivo* cancer imaging agent using conventional clinical 1.5 T MRI [86]. These particles have shown excellent dispersibility, low RES uptake, long-term stability under physiological condition, and low cytotoxicity, thus making poly(TMSMA-*r*-PEGMA)-coated SPION good candidates as MRI contrast agents for *in vivo* applications. Subcutaneous injection of the LLC cell line into the midback of mice generated tumors. After the intravenous injection of poly(TMSMA-*r*-PEGMA)-coated SPION, SW-SPION **16**, MRI of the mice was performed at different time points [86]. At 1 h postinjection, some areas of darkening of the $T_2$-weighted MR images were observed in the tumor area with a $T_2$ signal drop of 42%

which indicated the accumulation of detectable amounts of SPION within the tumor. At 4 h postinjection, 32% of the $T_2$ signal drop relative to that of the control was observed. The presence of the SPION in the tumor tissue was verified by Prussian blue staining. It is believed that the poly(TMSMA-*r*-PEGMA)-coated SPION accumulated in the tumor sites by EPR effect, due to the presence of leaky vasculature around the tumor, thus resulting in the efficient *in vivo* imaging of the cancer.

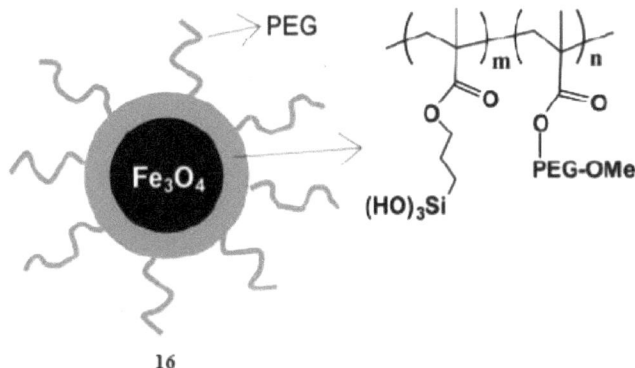

**Figure 22:** Poly(TMSMA-r-PEGMA)-coated superparamagnetic iron oxide nanoparticles (SPION). Reproduced with permission from Lee H, Lee E, Kim do K, Jang NK, Jeong YY, Jon S. Antibiofouling polymer-coated superparamagnetic iron oxide nanoparticles as potential magnetic resonance contrast agents for *in vivo* cancer imaging. J Am Chem Soc 2006; Jun 7, 128 (22): 7383-9.

Superparamagnetic magnetite nanoparticles were surface-modified with PEG and folic acid to obtain PEG-coated and folic acid-coated magnetite nanoparticles respectively [87]. It was found that the PEG- and folic acid-coated nanoparticles have higher uptake into human breast cancer cells (BT20) in comparison to the unmodified magnetite nanoparticles [87]. This suggests that the PEG or folic acid coating can resist the protein adsorption. As a result the macrophage cells fail to recognize the magnetite nanoparticles and the nanoparticles are taken up by the specific cancer cells (BT20) [87].

The surface charge of nanoparticles can play an important role in the uptake and biodistribution of nanoparticles both *in vitro* and *in vivo*. Usually the surface of nanoparticles is modified by different chemical moities such as PEG, amine, and carboxyl functionalities, which can alter surface charge of the nanoparticles. The positively charged nanoparticles can bind to cells due to electrostatic interaction with the negatively charged cell membranes, and then internalize into cells, while nanoparticles having negative surface charge may be taken up by the cells through protein-mediated phagocytosis and diffusion.

Albumin nanoparticles were coated with various primary amines to provide different surface charges. Under *in vitro* conditions, the albumin nanoparticles with zeta potential close to zero displayed much less phagocytic uptake than charged particles, especially nanoparticles with a positive zeta potential [88]. In contrast, the *in vivo* distribution of albumin nanoparticles in rats showed that various nanoparticles with positive, neutral and negative surface charges have similar blood circulation time and organ accumulation [88]. The results of this study suggest that *in vivo* conditions have additional factors affecting phagocytic uptake which are not reflected in an *in vitro* study.

Bakers and co-workers have synthesized the carboxyl-functionalized poly(amidoamine) (PAMAM) dendrimer-coated iron oxide nanoparticles [89]. The intracellular uptake of these dendrimer-stabilized nanoparticles was tested *in vitro* using KB cells (a human epithelial carcinoma cell line) which overexpress folate receptors. It was observed that the carboxyl terminated PAMAM dendrimer-stabilized iron oxide nanoparticles were taken up by KB cells despite the repulsive force between the negatively charged cells and the negatively charged nanoparticles [89]. In the presence of large amount of carboxyl terminal groups on the dendrimer surface, the folic acid-modified dendrimer coated iron oxide nanoparticles did not undergo receptor-mediated endocytosis [89]. The results of this study indicate that the surface charge of the dendrimer-stabilized magnetic iron oxide nanoparticles plays an important role in their biological activity.

The biodistribution of nanoparticles *in vivo* is affected by the size of the nanoparticles. Nanoparticles with diameters greater than 200 nm are usually taken up by the liver and spleen and eliminated by the Reticuloendothelial System (RES), resulting in decreased blood circulation times [90]; whereas smaller particles with diameter less than 5 nm are rapidly eliminated by the kidney [90, 91]. Thus, nanoparticles ranging from 5 nm to 100 nm can be optimal for use in targeted imaging and/or therapy, and offer the most effective biodistribution *in vivo*.

Chen and co-workers have studied the effect of MePEG molecular weight and particle size of TNF-α loaded stealth nanoparticles on *in vivo* tumor targeting [92]. MePEG of higher molecular weight provided greater fixed aqueous layer thickness (FALT) and particles of smaller size offered higher surface MePEG density. It was observed that serum protein adsorption and the phagocytic uptake were substantially decreased for nanoparticle of higher MePEG molecular weight or smaller particle size [92]. For example, nanoparticles made of $PEG_{5000}$-PHDCA (poly methoxypolyethylene glycol cyanoacrylate-co-n-hexadecyl cyanoacrylate) with a size of 80 nm, having a thicker FALT (5.16 nm) and a shorter distance (0.87 nm) between two neighbouring MePEG chains, exhibited the strongest ability to reduce protein adsorption and phagocytic uptake [92]. So, these particles increased the half-life of human TNF-α by 24-fold in S-180 tumor-bearing mice.

These studies provide new strategies to reduce the non-specific uptake of nanoparticles, prolong their blood half-life, and hence increase the target specificity and sensitivity for imaging as well as therapy [85].

## CONCLUSIONS

Recent progress in medicinal chemistry, genetics, molecular biology, material science, biomedical engineering, and nanotechnology have culminated in the development of superior probes for molecular imaging of cancer and therapeutic agents for targeted cancer therapy [3-8, 93]. Peptides, small molecules, peptidomimetics, and antibodies have been labeled with radioisotopes, fluorescent dyes, various nanoparticles, and quantum dots to develop probes for MRI, PET, SPECT, CT, and near IR-fluorescence imaging of various cancers [3-8, 93].

Although the progress in nanotechnology research has demonstrated the benefits of using nanoparticles for targeted tumor imaging and therapy, many of these nanotechnology-based receptor and antibody (or, peptide) targeted imaging and therapeutic agents reported in the literature are either 'proof of principles' or are only applicable *in vitro* and/or in small animal models, and are not suitable for translation into clinical use. Hence, it is imperative to develop nanoparticles that are biocompatible, highly target specific, give high signal to noise ratio in imaging, and have optimum pharmacokinetic and pharmacodynamic profiles without any undesired side effect. Such nanoparticles will be ideal for use as tumor-targeted imaging probes and therapeutic agents

To nurture the continued discovery and development of nanotechnology-based tumor-targeted imaging probes and therapeutic agents for targeted cancer therapy, concerted efforts are needed from molecular biologists to identify and validate suitable biomarkers for molecular imaging targets; medicinal chemists to design, synthesize and characterize target-specific imaging probes; biomedical engineers to develop high resolution, highly sensitive imaging devices having high signal-to-noise ratio; pharmaceutical scientists for pharmacokinetic and pharmacodynamic studies; and medical doctors for clinical applications. Only a collaborative research effort will lead to the discovery of clinically useful nanoparticles that can diagnose, prevent, and cure cancer. Nanotechnology-based targeted imaging and therapy of cancer will continue to evolve and play crucial role in modern-day cancer management.

## REFERENCES

[1]    Bradbury M, Hricak H. Molecular MR imaging in oncology. Magn Reson Imaging Clin N Am 2005;13(2):225-40.
[2]    Ito A, Kuga Y, Honda H *et al.* Magnetite nanoparticle-loaded anti-HER2 immunoliposomes for combination of antibody therapy with hyperthermia. Cancer Lett 2004; 212(2):167-75.

[3] Gambhir SS. Molecular imaging of cancer with positron emission tomography. Nat Rev Cancer 2002;2(9):683-93.

[4] Willmann JK, van Bruggen N, Dinkelborg LM *et al.* Molecular imaging in drug development. Nat Rev Drug Discov 2008;7(7):591-607.

[5] Weissleder R, Mahmood U. Molecular imaging. Radiology 2001;219(2):316-33.

[6] Massoud TF, Gambhir SS. Molecular imaging in living subjects: seeing fundamental biological processes in a new light. Genes Dev 2003;17(5):545-80.

[7] Hoffman JM, Gambhir SS. Molecular imaging: the vision and opportunity for radiology in the future. Radiology 2007;244(1):39-47.

[8] Peng XH, Qian X, Mao H *et al.* Targeted magnetic iron oxide nanoparticles for tumor imaging and therapy. Int J Nanomedicine 2008;3(3):311-21.

[9] Nie S, Xing Y, Kim GJ, Simons JW. Nanotechnology applications in cancer. Annu Rev Biomed Eng 2007;9:257-88.

[10] Atri M. New technologies and directed agents for applications of cancer imaging. J Clin Oncol 2006;24(20):3299-308.

[11] Pinkernelle J, Teichgraber U, Neumann F *et al.* Imaging of single human carcinoma cells *in vitro* using a clinical whole-body magnetic resonance scanner at 3.0 T. Magn Reson Med 2005;53(5):1187-92.

[12] Lucignani G. Nanoparticles for concurrent multimodality imaging and therapy: the dawn of new theragnostic synergies. Eur J Nucl Med Mol Imaging 2009;36(5):869-74.

[13] Del Vecchio S, Zannetti A, Fonti R *et al.* Nuclear imaging in cancer theranostics. Q J Nucl Med Mol Imaging 2007;51(2):152-63.

[14] Shubayev VI, Pisanic TR, Jin SH. Magnetic nanoparticles for theragnostics. Adva Drug Deliv Rev 2009;61(6):467-77.

[15] Pene F, Courtine E, Cariou A *et al.* Toward theragnostics. Critl Care Med 2009;37(1):S50-S8.

[16] Ozdemir V, Williams-Jones B, Glatt SJ, *et al.* Shifting emphasis from pharmacogenomics to theragnostics. Nature Biotechnol 2006;24(8):942-7.

[17] Wang X, Yang L, Chen ZG *et al.* Application of nanotechnology in cancer therapy and imaging. CA Cancer J Clin 2008 8(2):97-110.

[18] Balogh L, Nigavekar SS, Nair BM *et al.* Significant effect of size on the *in vivo* biodistribution of gold composite nanodevices in mouse tumor models. Nanomedicine 2007;3(4):281-96.

[19] Rapoport N, Gao Z, Kennedy A. Multifunctional nanoparticles for combining ultrasonic tumor imaging and targeted chemotherapy. J Natl Cancer Inst 2007;99(14):1095-106.

[20] Winter PM, Caruthers SD, Kassner A *et al.* Molecular imaging of angiogenesis in nascent Vx-2 rabbit tumors using a novel alpha(nu)beta3-targeted nanoparticle and 1.5 tesla magnetic resonance imaging. Cancer Res 2003;63(18):5838-43.

[21] Koo YE, Reddy GR, Bhojani M *et al.* Brain cancer diagnosis and therapy with nanoplatforms. Adva Drug Deliv Rev 2006;58(14):1556-77.

[22] Popovtzer R, Agrawal A, Kotov NA *et al.* Targeted gold nanoparticles enable molecular CT imaging of cancer. Nano Lett 2008;(12):4593-6.

[23] Anderson CJ, Ferdani R. Copper-64 radiopharmaceuticals for PET imaging of cancer: advances in preclinical and clinical research. Cancer Biother Radiopharm 2009;(4):379-93.

[24] Chen TJ, Cheng TH, Chen CY *et al.* Targeted Herceptin-dextran iron oxide nanoparticles for noninvasive imaging of HER2/neu receptors using MRI. J Biol Inorg Chem 2009;14(2):253-60.

[25] Cai W, Chen K, Li ZB *et al.* Dual-function probe for PET and near-infrared fluorescence imaging of tumor vasculature. J Nucl Med 2007;48(11):1862-70.

[26] Smith AM, Ruan G, Rhyner MN *et al.* Engineering luminescent quantum dots for *in vivo* molecular and cellular imaging. Ann Biomed Eng 2006;34(1):3-14.

[27] Gao X, Cui Y, Levenson RM *et al. In vivo* cancer targeting and imaging with semiconductor quantum dots. Nat Biotechnol 2004;22(8):969-76.

[28] Huang DH, Su L, Peng XH *et al.* Quantum dot-based quantification revealed differences in subcellular localization of EGFR and E-cadherin between EGFR-TKI sensitive and insensitive cancer cells. Nanotechnology 2009;20(22):225102.

[29] Akerman ME, Chan WC, Laakkonen P *et al.* Nanocrystal targeting *in vivo*. Proc Natl Acad Sci U S A 2002;99(20):12617-21.

[30]  Bagalkot V, Zhang L, Levy-Nissenbaum E *et al.* Quantum dot-aptamer conjugates for synchronous cancer imaging, therapy, and sensing of drug delivery based on bi-fluorescence resonance energy transfer. Nano Lett 2007;7(10):3065-70.

[31]  Cai W, Chen X. Preparation of peptide-conjugated quantum dots for tumor vasculature-targeted imaging. Nat Protoc 2008;3(1):89-96.

[32]  Cai W, Shin DW, Chen K *et al.* Peptide-labeled near-infrared quantum dots for imaging tumor vasculature in living subjects. Nano Lett 2006;6(4):669-76.

[33]  Chen K, Li ZB, Wang H *et al.* Dual-modality optical and positron emission tomography imaging of vascular endothelial growth factor receptor on tumor vasculature using quantum dots. Eur J Nucl Med Mol Imaging 2008;35(12):2235-44.

[34]  Yang L, Mao H, Wang YA *et al.* Single chain epidermal growth factor receptor antibody conjugated nanoparticles for *in vivo* tumor targeting and imaging. Small 2009;5(2):235-43.

[35]  Smith AM, Duan H, Mohs AM *et al.* Bioconjugated quantum dots for *in vivo* molecular and cellular imaging. Adv Drug Deliv Rev 2008;60(11):1226-40.

[36]  Yu X, Chen L, Li K *et al.* Immunofluorescence detection with quantum dot bioconjugates for hepatoma *in vivo*. J Biomed Opt 2007;12(1):014008.

[37]  Stroh M, Zimmer JP, Duda DG *et al.* Quantum dots spectrally distinguish multiple species within the tumor milieu *in vivo*. Nat Med 2005;11(6):678-82.

[38]  Gao J, Chen K, Xie R *et al.* Ultrasmall Near-Infrared Non-cadmium Quantum Dots for *in vivo* Tumor Imaging. Small (Weinheim an der Bergstrasse, Germany). 2009 Nov 12.

[39]  Kim S, Lim YT, Soltesz EG *et al.* Near-infrared fluorescent type II quantum dots for sentinel lymph node mapping. Nat Biotechnol 2004;22(1):93-7.

[40]  Hardman R. A toxicologic review of quantum dots: toxicity depends on physicochemical and environmental factors. Environment Health Perspect 2006;114(2):165-72.

[41]  El-Sayed I, Huang X, Macheret F *et al.* Effect of plasmonic gold nanoparticles on benign and malignant cellular autofluorescence: a novel probe for fluorescence based detection of cancer. Technol Cancer Res Treat 2007;6(5):403-12.

[42]  Jackson PA, Rahman WN, Wong CJ *et al.* Potential dependent superiority of gold nanoparticles in comparison to iodinated contrast agents. Eur J Radiol 2009 Apr 28.

[43]  Reck M, Crino L. Advances in anti-VEGF and anti-EGFR therapy for advanced non-small cell lung cancer. Lung Cancer 2009;63(1):1-9.

[44]  Lafky JM, Wilken JA, Baron AT *et al.* Clinical implications of the ErbB/epidermal growth factor (EGF) receptor family and its ligands in ovarian cancer. Biochim Biophys Acta 2008;1785(2):232-65.

[45]  Reuter CW, Morgan MA, Eckardt A. Targeting EGF-receptor-signalling in squamous cell carcinomas of the head and neck. Br J Cancer 2007;96(3):408-16.

[46]  Nicholson RI, Gee JM, Harper ME. EGFR and cancer prognosis. Eur J Cancer 2001;37 Suppl 4:S9-15.

[47]  Chrysogelos SA, Dickson RB. EGF receptor expression, regulation, and function in breast cancer. Breast Cancer Res Treat 1994;29(1):29-40.

[48]  Huang X, El-Sayed IH, Qian W *et al.* Cancer cell imaging and photothermal therapy in the near-infrared region by using gold nanorods. J Am Chem Soc 2006;128(6):2115-20.

[49]  Sokolov K, Follen M, Aaron J *et al.* Real-time vital optical imaging of precancer using anti-epidermal growth factor receptor antibodies conjugated to gold nanoparticles. Cancer Res 2003;63(9):1999-2004.

[50]  Qian X, Peng XH, Ansari DO *et al. In vivo* tumor targeting and spectroscopic detection with surface-enhanced Raman nanoparticle tags. Nat Biotechnol 2008;26(1):83-90.

[51]  Pernodet N, Fang X, Sun Y *et al.* Adverse effects of citrate/gold nanoparticles on human dermal fibroblasts. Small (Weinheim an der Bergstrasse, Germany) 2006;2(6):766-73.

[52]  Cho WS, Cho M, Jeong J *et al.* Acute toxicity and pharmacokinetics of 13 nm-sized PEG-coated gold nanoparticles. Toxicol App Pharmacol 2009;236(1):16-24.

[53]  Caravan P, Ellison JJ, McMurry TJ *et al.* Gadolinium(III) Chelates as MRI Contrast Agents: Structure, Dynamics, and Applications. Chem Rev 1999;99(9):2293-352.

[54]  Ito H. Oncology imaging: current status and new approaches. Intern J Clin Oncol / Japan Soc Clin Oncol 2006;11(4):256-7.

[55]  Cai W, Chen X. Multimodality molecular imaging of tumor angiogenesis. J Nucl Med 2008;49 Suppl 2:113S-28S.

[56] Hu G, Lijowski M, Zhang H *et al*. Imaging of Vx-2 rabbit tumors with alpha(nu)beta3-integrin-targeted 111In nanoparticles. Int J Cancer 2007;120(9):1951-7.

[57] Lijowski M, Caruthers S, Hu G *et al*. High sensitivity: high-resolution SPECT-CT/MR molecular imaging of angiogenesis in the Vx2 model. Invest Radiol 2009;44(1):15-22.

[58] Harisinghani MG, Barentsz J, Hahn PF *et al*. Noninvasive detection of clinically occult lymph-node metastases in prostate cancer. N Engl J Med 2003;348(25):2491-9.

[59] Funovics MA, Kapeller B, Hoeller C *et al*. MR imaging of the her2/neu and 9.2.27 tumor antigens using immunospecific contrast agents. Magn Reson Imaging 2004;22(6):843-50.

[60] Jain TK, Morales MA, Sahoo SK *et al*. Iron oxide nanoparticles for sustained delivery of anticancer agents. Mol Pharm 2005;2(3):194-205.

[61] Montet X, Montet-Abou K, Reynolds F *et al*. Nanoparticle imaging of integrins on tumor cells. Neoplasia 2006;8(3):214-22.

[62] Thorek DL, Chen AK, Czupryna J *et al*. Superparamagnetic iron oxide nanoparticle probes for molecular imaging. Ann Biomed Eng 2006;34(1):23-38.

[63] Cerdan S, Lotscher HR, Kunnecke B *et al*. Monoclonal antibody-coated magnetite particles as contrast agents in magnetic resonance imaging of tumors. Magn Reson Med 1989;12(2):151-63.

[64] Sipkins DA, Cheresh DA, Kazemi MR *et al*. Detection of tumor angiogenesis *in vivo* by alphaVbeta3-targeted magnetic resonance imaging. Nat Med 1998;4(5):623-6.

[65] Storrs RW, Tropper FD, Li HY *et al*. Paramagnetic polymerized liposomes as new recirculating MR contrast agents. J Magn Reson Imaging 1995;5(6):719-24.

[66] Artemov D, Mori N, Okollie B *et al*. MR molecular imaging of the Her-2/neu receptor in breast cancer cells using targeted iron oxide nanoparticles. Magn Reson Med 2003;49(3):403-8.

[67] Huh YM, Jun YW, Song HT *et al*. *In vivo* magnetic resonance detection of cancer by using multifunctional magnetic nanocrystals. J Am Chem Soc 2005;127(35):12387-91.

[68] Yang L, Mao H, Cao Z *et al*. Molecular imaging of pancreatic cancer in an animal model using targeted multifunctional nanoparticles. Gastroenterology 2009;136(5):1514-25 e2.

[69] Yang L, Peng XH, Wang YA *et al*. Receptor-targeted nanoparticles for *in vivo* imaging of breast cancer. Clin Cancer Res 2009;15(14):4722-32.

[70] Moore A, Medarova Z, Potthast A *et al*. *In vivo* targeting of underglycosylated MUC-1 tumor antigen using a multimodal imaging probe. Cancer Res 2004;64(5):1821-7.

[71] Leuschner C, Kumar CS, Hansel W *et al*. LHRH-conjugated magnetic iron oxide nanoparticles for detection of breast cancer metastases. Breast Cancer Res Treat 2006;99(2):163-76.

[72] Zhang C, Jugold M, Woenne EC *et al*. Specific targeting of tumor angiogenesis by RGD-conjugated ultrasmall superparamagnetic iron oxide particles using a clinical 1.5-T magnetic resonance scanner. Cancer Res 2007;67(4):1555-62.

[73] Simberg D, Duza T, Park JH *et al*. Biomimetic amplification of nanoparticle homing to tumors. Proc Natl Acad Sci U S A 2007;104(3):932-6.

[74] Low PS, Henne WA, Doorneweerd DD. Discovery and development of folic-Acid-based receptor targeting for imaging and therapy of cancer and inflammatory diseases. Acc Chem Res 2008;41(1):120-9.

[75] Sun C, Sze R, Zhang M. Folic acid-PEG conjugated superparamagnetic nanoparticles for targeted cellular uptake and detection by MRI. J Biomed Mater Res A 2006;78(3):550-7.

[76] Chen H, Gu Y, Hub Y *et al*. Characterization of pH- and temperature-sensitive hydrogel nanoparticles for controlled drug release. PDA J Pharm Sci Technol 2007;61(4):303-13.

[77] Devaraj NK, Keliher EJ, Thurber GM *et al*. 18F labeled nanoparticles for *in vivo* PET-CT imaging. Bioconjug Chem 2009;20(2):397-401.

[78] Montet X, Weissleder R, Josephson L. Imaging pancreatic cancer with a peptide-nanoparticle conjugate targeted to normal pancreas. Bioconjug Chem 2006;17(4):905-11.

[79] Santra S, Kaittanis C, Grimm J *et al*. Drug/dye-loaded, multifunctional iron oxide nanoparticles for combined targeted cancer therapy and dual optical/magnetic resonance imaging. Small 2009;5(16):1862-8.

[80] Pan D, Caruthers SD, Hu G *et al*. Ligand-directed nanobialys as theranostic agent for drug delivery and manganese-based magnetic resonance imaging of vascular targets. J Am Chem Soc 2008;130(29):9186-7.

[81] Rowe MD, Thamm DH, Kraft SL *et al*. Polymer-modified gadolinium metal-organic framework nanoparticles used as multifunctional nanomedicines for the targeted imaging and treatment of cancer. Biomacromolecules 2009;10(4):983-93.

[82] Schmieder AH, Caruthers SD, Zhang H *et al.* Three-dimensional MR mapping of angiogenesis with alpha5beta1(alpha nu beta3)-targeted theranostic nanoparticles in the MDA-MB-435 xenograft mouse model. FASEB J 2008;22(12):4179-89.

[83] Moore A, Weissleder R, Bogdanov A, Jr. Uptake of dextran-coated monocrystalline iron oxides in tumor cells and macrophages. J Magn Reson Imaging 1997;7(6):1140-5.

[84] Lee JH, Huh YM, Jun YW *et al.* Artificially engineered magnetic nanoparticles for ultra-sensitive molecular imaging. Nat Med 2007;13(1):95-9.

[85] Rogers WJ, Basu P. Factors regulating macrophage endocytosis of nanoparticles: implications for targeted magnetic resonance plaque imaging. Atherosclerosis 2005;178(1):67-73.

[86] Lee H, Lee E, Kim do K *et al.* Antibiofouling polymer-coated superparamagnetic iron oxide nanoparticles as potential magnetic resonance contrast agents for *in vivo* cancer imaging. J Am Chem Soc 2006;128(22):7383-9.

[87] Zhang Y, Kohler N, Zhang M. Surface modification of superparamagnetic magnetite nanoparticles and their intracellular uptake. Biomaterials 2002;23(7):1553-61.

[88] Roser M, Fischer D, Kissel T. Surface-modified biodegradable albumin nano- and microspheres. II: effect of surface charges on *in vitro* phagocytosis and biodistribution in rats. Eur J Pharm Biopharm 1998;46(3):255-63.

[89] Shi X, Thomas TP, Myc LA *et al.* Synthesis, characterization, and intracellular uptake of carboxyl-terminated poly(amidoamine) dendrimer-stabilized iron oxide nanoparticles. Phys Chem Chem Phys 2007;9(42):5712-20.

[90] Gupta AK, Gupta M. Synthesis and surface engineering of iron oxide nanoparticles for biomedical applications. Biomaterials 2005;26(18):3995-4021.

[91] Choi HS, Liu W, Misra P *et al.* Renal clearance of quantum dots. Nat Biotechnol 2007;25(10):1165-70.

[92] Fang C, Shi B, Pei YY *et al. In vivo* tumor targeting of tumor necrosis factor-alpha-loaded stealth nanoparticles: effect of MePEG molecular weight and particle size. Eur J Pharm Sci 2006;27(1):27-36.

[93] Philip R, Murthy S, Krakover J *et al.* Shared immunoproteome for ovarian cancer diagnostics and immunotherapy: potential theranostic approach to cancer. J Proteome Res 2007;6(7):2509-17.

# CHAPTER 6

# The Nanotoxicological Influence of Nanoparticles, *with Special Reference to the Vasculature*

## May Azzawi[*]

*Manchester Metropolitan University, Manchester, UK*

**Abstract:** In the last decade, there has been a dramatic increase in the synthesis of engineered nanomaterials for a wide variety of applications. Despite the clear advantages of these applications, especially in medical intervention and therapeutics, the influence of nanoparticle exposure on cellular and organ function remains poorly understood. The nanoscale size of nanoparticles (<100 nm) allows their penetration into cells and, due to their large surface area per unit mass, they are more reactive than larger scale particles. This has led government and scientific organisations to call for a need to assess the safety of engineered nanomaterials and determine the mechanism of their interaction with cells and tissues. Nanotoxicity is now an emerging and expanding discipline that addresses these issues to ensure the well being of living organisms. The results from recent studies in the literature demonstrate that the nanotoxicological influence of nanoparticles is dependent on their size, charge, cell type and material composition. There also remains a number of methodological considerations and the need for precise physicochemical characterisation of nanoparticles before their use in nanotoxicity studies. Recent studies also highlight the need to adequately assess the toxic effects of nanomedicines on all cells and tissues, including the vasculature, using *in vitro* as well as *in vivo* studies.

**Keywords:** Nanomaterial; nanotoxicity; vasculature; uptake; endocytosis; smooth muscle cells; endothelial cells, caveolae; contractility.

## INTRODUCTION

The continued emergence of newly synthesized nanomaterials and the fabrication of nanoparticles with novel properties has meant that there are ever increasing prospects for new applications, especially in the field of medicine. Nanomaterials (<100 nm) used for medical intervention and therapeutics are collectively termed *'Nanomedicines'* [1]. Examples include the use of drug-loaded and/or labelled nanoparticles for a wide number of disease conditions and in imaging diagnostics [2-4]. The efficacy of these applications has been examined *in vivo* using animal models of disease [2, 4] as well as *in vitro*, using cell cultures [5, 6]. Findings from these studies show that although the use of encapsulated or tagged nanoparticles minimises systemic drug toxicity and reduces drug dosage [2, 7], their influence on cellular function remains poorly understood.

Besides their use in medical intervention and therapeutics, the wide availability of nanoparticles in every day applications (chemical, domestic, cosmetic, etc), has meant that our bodies are continually exposed to nanoparticles *via* various routes of entry, through inhalation, digestion and skin contact. At the nanoscale, both physical and chemical properties of materials can differ significantly from those at the larger scale. Nanoparticles have a larger surface area per unit mass than larger scale particles and hence can be more reactive with the surrounding biological tissues and cell components. They can also more readily penetrate cells and tissues. *In vivo* nanoparticle exposure studies have shown that nanoparticles can gain entry into the blood stream, leading to their systemic uptake where they can be detected within organs that are distal to the exposure site [8-10]. Evidence suggests that nanoparticles can be taken up by non-phagocytic cells, such as endothelial cells where they enter the cytoplasm and can interact with cellular components. This highlights a major concern relating to the influence of nanoparticles on cellular and organ function.

*Addres correspondence to May Azzawi: School of Healthcare Science, Faculty of Science and Engineering,Manchester Metropolitan University, Manchester, UK; Tel: (0161) 2473332; E-mail: m.azzawi@mmu.ac.uk

**Mark Slevin (Ed)**

Consequently, it has become necessary to consider the health and safety implications of the use of new nanotechnologies, including nanomedicines. Nanotoxicology is thus an emerging and expanding discipline, defined as *'the study of the adverse effects of engineered nanomaterials on living organisms and the ecosystems, including the prevention and amelioration of such adverse effects'* [1]. Both government [11] and scientific organisations [12] have called for a need to assess the safety of engineered nanomaterials. The recommendations are that studies must be carried out into the mechanisms of interactions of nanoparticles with cells and their components and the toxic affects on tissues, in particular the heart, skin, nervous system and blood vessels [12]. The role of blood vessels is often overlooked in nanotoxicological studies. The vascular tree, comprising major conduit vessels, arteries, veins and the microvasculature, supply blood to all tissues and organs to maintain health and well being. An essential characteristic of the vasculature is the ability to control the rate of blood flow through a number of mechanisms, including the regulation of vessel diameter to ensure that flow is matched to tissue demand. The innermost endothelial layer lining all blood vessels plays an important role in the modulation of vessel diameter. It also plays a critical role in maintaining vascular homeostasis through the release of a number of mediators. The uptake of nanoparticles by endothelial cells may disturb homeostatic mechanisms leading to 'endothelial dysfunction' which is a hallmark of many conditions, including diabetes, peripheral vascular disease, atherosclerosis and restenosis.

This chapter will review studies from the literature investigating the nanotoxicological influence of nanomaterials on cells and tissues, in particular, the non-phagocytic cells of the vasculature. The mode of nanoparticle exposure and their penetration into the multiple layers of the vessel wall will be explored. Finally, studies investigating the effects of nanoparticle uptake on cellular and vascular function (including vasodilator and vasoconstrictor responses) will be addressed. Emphasis will be made on the effects of nanomaterials used for drug-loading and imaging diagnostics, using silica and gold nanoparticles as model examples.

## 1. NANOTOXICITY OF NANOPARTICLES

### 1.1 Nanoparticle Uptake

Nanoparticle uptake into cells and tissues has been studied using *in vitro* cell cultures and subsequent to intravenous administration in whole animals. In particular, the mechanism of nanoparticle uptake by non-phagocytic cells such as endothelial cells and vascular smooth muscle cells has attracted attention because they are potential targets for therapeutic intervention in the treatment of vascular disease. Evidence suggests that nanoparticles can be taken up into cells by specific as well as non-specific processes, depending on their size, shape and surface property.

Studies using vascular smooth muscle cell cultures show that nanoparticle uptake is an energy dependent process. When vascular smooth muscle cells were incubated with biodegradable polymeric nanoparticles (approx. 150 nm diameter), uptake at 37°C reached saturation after 12 hours of incubation, however, at 4°C, cellular uptake was reduced to 10% [13]. Furthermore, the work by Labhasetwar's group has demonstrated that nanoparticle uptake is also a dynamic process whereby nanoparticles are taken up in a time- and concentration- dependent manner. In one study, the uptake of biodegradable nanoparticles (*PLGA*- poly D,L-lactide-co-glycolide, 97 ± 3 nm) by human arterial vascular smooth muscle cells *in vitro*, and shown to be inhibited in the presence of metabolic inhibitors. The uptake of nanoparticles was concentration dependent being linear at lower doses of 10-100 µg, but reduced efficiency of uptake at higher doses of 500-1000 µg. Nanoparticles were internalised into cells within 1 minute of incubation, reaching a saturation uptake point within only 4-6 hours. Their findings suggest that the mechanism of uptake is *via* an endocytic process that involves both fluid phase pinocytosis and non-specific receptor-mediated endocytosis, but not by phagocytosis [14]. Furthermore, they were also able to demonstrate that a reduction in intracellular nanoparticle retention was due to their expulsion from the cell by an energy dependent process of exocytosis [14]. Nanoparticle uptake into human endothelial cells (human umbilical vein endothelial cells) has also been shown to be time and concentration dependent [15], where nanoparticles are taken up into endosomes and subsequently transported to lysosomes. They are then able to escape endolysosomal entrapment and move into the cytoplasm [16]. In a study exploring the size dependence of endocytic particle uptake, an upper limit of 200nm size was evident for fluorescent latex particle uptake (size range 50-1000 nm) into a melanoma cell line. Internalization was *via* clathrin-mediated rather than caveolae-mediated endocytosis [17].

In studies examining the size dependence of gold nanoparticle uptake, the highest degree of cellular uptake was for the 50 nm gold nanoparticle size (for diameter range 14-100 nm) with an uptake half life of 1.90 hours. The uptake was also dependent on the physical dimensions of nanoparticles being greater for spherical rather than rod shaped nanoparticles [18, 19]. Cellular uptake was enhanced by a thousand fold when gold nanoparticles were incorporated in a liposomal carrier [20]. Other uptake mechanisms have also been reported for smaller size gold nanoparticles. Geiser and colleagues [21] were able to show that uptake of gold nanoparticle (25 nm) at $6.6 \times 10^{10}$ particles /ml into red blood cells (24 hours, 37 °C) occurred by diffusion or by electrostatic, van der Waals or stearic interactions. Nanoparticles were not membrane bound, but freely localised [21].

An important consideration in assessing the influence of nanoparticle size and charge on cellular uptake is the fact that this can vary depending on their surroundings and the medium in which they are suspended. Nanoparticles have a large surface area to volume ratio and depending on their material composition and surface charge, they can attract proteins and /or lipids from their surroundings (in a dynamic process of adsorption and desorption) thus acquiring a secondary coating or 'corona' (reviewed in [1]). This provides nanoparticles with additional surface characteristics that can affect their uptake into cells. Investigators have utilised this feature to allow enhanced uptake into cells, as illustrated in the use of the protein albumin to enhance nanoparticle uptake into endothelial cells [22]. In addition to altered cellular uptake and cytotoxicity, nanoparticle suspension in protein containing solutions (such as physiological media) can also lead to altered zeta potential and charge, and an increase in nanoparticle size [23] and behaviour (including agglomeration). For example, in a study examining nanoparticle uptake into endothelial cells, detailed characterisation of the commercially supplied alumina nanoparticle of 10-20 nm diameter, using TEM and dynamic scatter analysis, demonstrated the additional presence of particle agglomerates of greater than 500nm when suspended in serum containing media [24]. These properties are often overlooked in many studies examining the influence of nanoparticle size and characteristics on cellular function [24, 25]. Determining the precise physico-chemical characteristics of nanoparticles within the suspended media is therefore paramount to the interpretation of findings.

## 1.2 Cytotoxicity of Nanoparticles

*In vivo* animal studies examining the toxicity, tissue accumulation and excretion of nanoparticles subsequent to their intravenous injection, have demonstrated that nanoparticles can influence cellular function and viability, induce an inflammatory response, and also stimulate oxidative stress pathways. These responses have been detected for nanoparticles of various material composition, including silica [8, 26], polystyrene [23] and metal and metal oxides [9, 10], but they are influenced by nanoparticle size, shape and the particular cell type. For example, a single injection of silica nanoparticles into mice (at 50 mg/kg) has been shown to lead to upregulation of cytokine mRNA and protein levels within the blood and intraperitoneal macrophages (IL-1β and TNF-α) [26]. Inflammatory responses were significant for 100 and 200 nm size silica nanoparticles, but not those of 50 nm nanoparticle [8]. Silica nanoparticles were seen within macrophages in the spleen and liver after 4 weeks of nanoparticle injection [8].

In the case of gold nanoparticles, a similar inflammatory response is observed. When injected intravenously into male BALB/c mice, PEG-coated gold nanoparticles have been shown to induce acute inflammation and apoptosis in the liver [27]. The study demonstrated a significant increase in cytokine (IL-1β, IL-6, IL-10, TNF-α) and chemokine (CCL-3, CCL-4) mRNA expression within 30 minutes of administration. Messenger RNA levels for adhesion molecules (ICAM-1 and E-Selectin) were significantly increased within 4 hours of administration. Increases were dose dependent. The number of apoptotic cells in the liver was significantly increased after 7 days of administration. PEG-coated gold nanoparticles accumulated in the vesicles and lysosomes of liver kupfer cells and spleen macrophages, in a time dependent manner. There was no evidence of nanoparticle translocation into the nucleus, mitochondria or the golgi complex, nor in tissues such as the brain, kidney, testes and lung [27]. However, the biodistribution of gold nanoparticle into tissues after intravenous administration have been shown to be particle size dependent [9]. Once injected intravenously, nanoparticles can exert their effects over a given period of time. Indeed, two phases of toxicity, induced by nanoparticles have been described: an immediate acute phase, observed subsequent to injection of the nanoparticles; and a second delayed phase, once the nanoparticles have accumulated in the tissues, often observed after 7 days [27]. Perhaps the most pronounced *in vivo* cytotoxic effects have been observed subsequent to administration of unmodified mesoporous silica nanoparticles (MSN). Surprisingly, intravenous and intraperitonial (but not subcutaneous) injection of MSN (MCM-41, size 150 nm), lead to severe systemic toxicity and death in mice (at concentrations above 5 mg/ml), possibly due to thrombus formation [28].

In terms of acute cellular effects, *in vitro* cell culture studies have examined the influence of nanoparticle uptake on cell survival and function by assessing effects on cell viability (cell lysis and death; apoptosis and necrosis), cell damage (cell and organelle disruption at the ultrastructural level), cell proliferation and migration as well as the influence on gene and protein expression of molecules within the cell. Effects on the expression of inflammatory mediators and the generation of free radicals (leading to oxidative stress) have attracted particular attention.

The use of both silica and polymeric nanoparticles has been explored as vehicles for drug delivery due to their reported biocompatible nature. However, evidence suggests a concentration and size dependent effect that is cell specific. While dye-loaded PLGA nanoparticles (250-300 nm) had no significant effect on the viability of human vascular smooth muscle cells [29], uptake of silica nanoparticles ($50\pm3$ nm) affected the survival rate of cultured lung epithelial cells in a concentration dependent manner. No significant reduction in cell proliferation rate or cell death, for nanoparticle concentrations up to 0.1 mg /ml, however, at 0.5 mg/ml (equivalent to $10^8 - 5\times10^{11}$ nanoparticle /ml) a significant reduction in survival rate was apparent [30]. Mesoporous silica (MSN) nanoparticles have gained special interest because of their potential use as drug loaded nanoparticles. These structures are porous with a hexagonal exterior, such that a range of shapes is possible thus providing different aspect ratios and pore sizes [31]. Huang and colleagues [32] examined the effect of MSN nanoparticle shape on cellular function and demonstrated that nanoparticles with greater aspect ratio (hence greater contact with intracellular components) were internalised faster into human melanoma cells and had bigger effects on cellular function than those with smaller aspect ratio. For instance, they led to increased cell adhesion and migration, but had no significant effect on cell viability (apoptosis) [32]. Spherical MSN caused no disruption to the cell cytoskeleton while long rod-shaped ones, did disrupt cytoskeletal organisation [32]. Cytotoxic effects of MSN were also observed in mouse myoblasts and mesothelial cells after a 24 hour culture period, but were shown to be dependant on concentration, particle and pore size [28]. In a human colon carcinoma cell line, however, cytotoxic effects were observed within only 3 hours of incubation and lead to increased ROS generation and increased apoptotic signalling [33].

A similar influence on cellular function has been demonstrated for metal and metal oxides, although this varies significantly depending on the material composition of nanoparticles. For example, uptake of both yttrium oxide and zinc oxide (but not iron oxide) nanoparticles by human aortic endothelial cells has been shown to lead to upregualtion of mRNA and protein expression of adhesion molecules and inflammatory mediators ( IL-8 and monocyte chemotactic protein-1) [34] and the generation of ROS [35]. Upregulation of adhesion molecules by porcine pulmonary artery endothelial cells and human umbilical vein endothelial cells has also been observed subsequent to incubation in alumina oxide nanoparticles (10-20 nm) [24]. Despite claims that gold nanoparticles are non-toxic [36, 37], some studies suggest that they can have a cytotoxic influence on cell integrity and function. For example, gold nanoparticle (10 nm) uptake into granulosa cells has been shown to alter organelle ultrastructure, such as mitochondria, and alter oestradiol secretion in a time dependent manner within a 24 hour period [38]. Uptake of gold nanoparticles into human dermal fibroblasts also led to cytoskeletal filament disruption and increased apoptosis in a size, time and concentration dependent manner [39]. In retinal cells, however, gold nanoparticle uptake had no effect on cell viability. Cultured human retinal microvascular endothelial cells were incubated in 20 nm and 100 nm gold nanoparticles at concentrations up to100 µM/L, over a 48 hour period. Nanoparticles were found bound on the membrane of cells but had no effect on the cell's viability or ultrastructure [37]. These findings highlight the influence of material composition as well as cell type specificity in the cellular responses to nanoparticles.

## 1.3 Mechanisms Leading to Cytotoxicity

Given their nanosize, and their ability to interact with intracellular protein components, the effects of nanoparticles on signal transduction mechanisms have recently been investigated [5, 40]. Evidence suggests that once nanoparticles are endocytosed, they can indeed interfere with signal transduction pathways, whereby they can uncouple and activate enzymes leading to the release of signalling molecules and mediators. For example, uptake of polysiloxane nanoparticles ($66\pm 30$ nm) by human aortic endothelial cells has been shown to induce nitric oxide release, *via* activation of the caveolae associated endothelial nitric oxide synthase (eNOS) enzymes [5]. In the case of gold nanoparticles, gold nanoparticle-polyurethane nanocomposites have been demonstrated to induce endothelial cell proliferation and migration by acting *via* VEGF-R2 and activation of FAK (focal adhesion kinase) and the P13K/Akt signalling pathway leading to eNOS activation and nitric oxide release [40]. Conversely, superoxide generation (*via* JNK activation) and upregulation of hemeoxygenase mRNA, has been documented in human aortic endothelial cells, following nanoparticle exposure (ultrafine particles from diesel exhaust) [41].

Dual concentration dependent effects have also been observed in the case of silver nanoparticles. At high concentrations (50-100 µg/ml) they have been shown to stimulate coronary endothelial cell proliferation *in vitro*, through activation of eNOS (*via* phosphorylation of Serine 1177). At low concentrations (below 10 µg/ml) however, they inhibited proliferation [25]. The dual effect on cell proliferation, documented by the authors, may be due to nanoparticle size heterogeneity (size range 10-90 nm with a tendency to aggregate).

A unique material worthy of note is cerium oxide. Exposure of human aortic endothelial cells to cerium oxide nanoparticles led to minimal inflammation, in comparison to other metal oxides [35, 42]. Indeed, recent studies have evaluated the effect of cerium oxide nanoparticles on retinal endothelial cell function [6] and demonstrated that they can act as free radical scavengers (size range 1-20 nm). The nanoparticles may represent a novel therapeutic tool for the treatment of conditions where reduced vasodilator responses are seen.

## 2. NANOPARTICLE PENETRATION INTO THE VASCULATURE

### 2.1 Vascular Exposure to Nanoparticles

Nanoparticles can penetrate blood vessels *via* two modes of exposure: Intravascular and Extravascular (see Fig. **1**). Intravascular (also termed intralumenal) exposure of blood vessels is achieved mainly through intravenous, {as well as intra-arterial [43]} injection of 'engineered' nanoparticles for drug delivery, medical imaging, or as medical device [3, 7]. Nanoparticles come in direct contact with the endothelial cell layer lining the vessel wall where they can be taken up either actively or passively. In the case of active uptake, this is achieved through tagging nanoparticles with endothelial specific ligands [7]. Nanoparticle suspensions can also be infused into the target tissue or organ, using infusion catheters, which are inserted into the lumen of blood vessels. Additionally, drug-eluting stents (coated with drug-loaded nanoparticles or polymeric coatings) can be inserted into the lumen of large and medium size arteries. This allows for localised, implant-based drug delivery, sometimes through perfusion holes through the vessel walls to prevent restenosis.

Extravascular exposure of blood vessels to nanoparticles is achieved through extracellular tissues and interstitial spaces, where nanoparticles come in contact with the outer adventitial layer of blood vessels. Nanoparticles translocating through this route include those that are absorbed through the mucosa *via* inhalation, ingestion or *via* skin absorption [44, 45]. These include pollutants and components of chemical, food and cosmetic reagents. High concentrations of intravenously, intradermally and intraperitoneally injected nanoparticles also translocate into the extracellular spaces and come in contact with other surrounding blood and lymph vessels, where they can penetrate through the entire vessel walls *via* the adventitial layer and reaching into the endothelial cell layer. Evidence suggests that indeed nanoparticle uptake *via* the GI track and the dermis can lead to nanoparticle localisation in the blood [44] and in nearby lymph nodes [46].

Intra-arterial infusion studies [47] demonstrate that in addition to vascular uptake into the media and adventitia, nanoparticle deposition is also seen in the myocardium, liver and lung. However, Oberdorster [1] argues that the biodistribution of intravenously administered nanoparticle is different from that of nanoparticles entering the blood *via* other routes of entry such as the respiratory track and that translocation of inhaled nanoparticles into the blood is slow, involving much smaller amounts of nanoparticles than originally administered [1]. Nanoparticle translocation across alveolar epithelial cells and capillary endothelial cells into the circulatory system can take place *via* different forms of caveolae, through capillary endothelial small and large fenestrae and intercellular gaps. In disease states they can also move across widened tight junctions {Reviewed in [48]}.

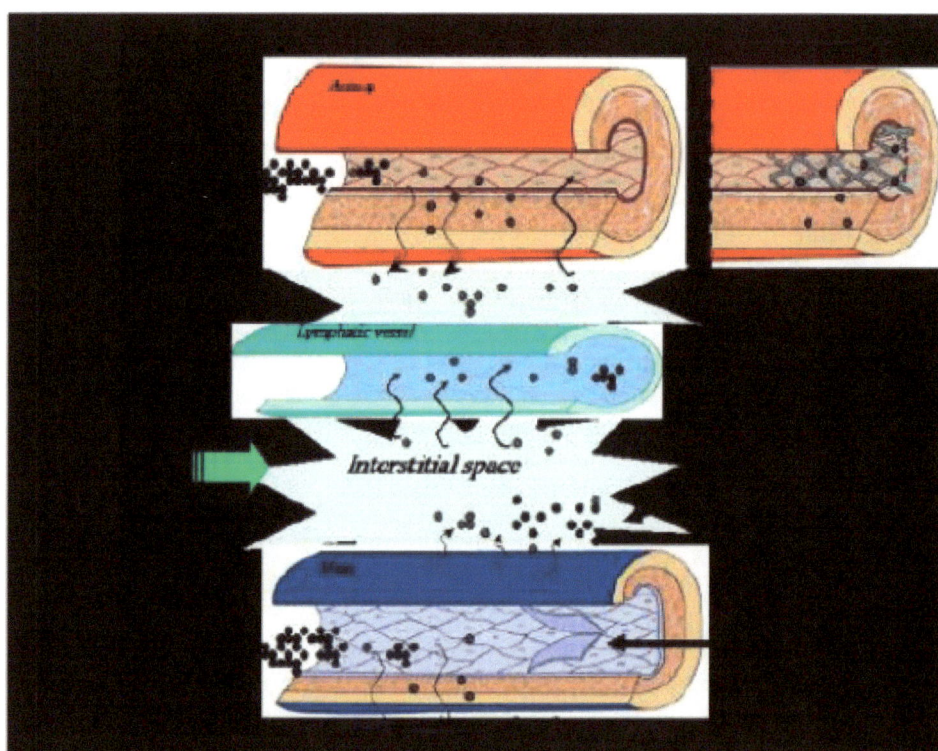

**Figure 1:** Schematic diagram illustrating two modes of vascular exposure to nanoparticle. Intravascular exposure is shown *via* intravenous and intra-arterial injection; and stent implantation. Extravascular exposure is shown *via* nanoparticle penetration through the vasculature and translocation into the interstitial spaces.

## 2.2 Nanoparticle Retention in Blood

An ideal property of nanoparticles for drug delivery includes 'long circulating half-time', where longer retention time in the blood is associated with improved efficacy of nanomedicines. Furthermore, nanoparticles used for imaging as contrast agents also require long blood circulation times. The blood half life of nanoparticles depends on the material composition, size and shape of nanoparticles as well as the concentration of the administered nanoparticles. For example, the blood half life for 13 nm PEG-coated gold nanoparticles is 32.65 ±11.64 and 28.50 ± 4.09 hours when administered at 4.26 and 0.85 mg/kg in BALB/c mice [27]. This contrasts with a shorter half life of 14.5 ± 3.27 hours in BALB/c mice for the larger 38 nm PEG-coated gold nanoparticles [49]. In contrast, the blood half life for other metal oxides, such as superparamagnetic iron oxide nanoparticles, is only 0.27 hours in Sprague-Dawley rats [50]. Long circulation half time of nanoparticles, therefore, poses the problem that blood vessels are exposed intravascularly to nanoparticles over a given period of time, particularly, exposure of the lining endothelial cell layer. This is compounded by the fact that enhanced arterial uptake is necessary to achieve efficient drug delivery into the target tissues, which leads to further nanoparticle exposure of the inner medial and adventitial layers of the vascular wall.

## 2.3 Nanoparticle Uptake into the Vasculature

There is much interest in the penetration properties of drug loaded nanoparticle from nanoparticle- coated stents and localised catheter based delivery. Studies demonstrate that uptake of biodegradeable polymers is particle size dependent, with 100-200 nm particles showing better penetration than 514 nm size. Penetration was also enhanced by use of pressure-induced infusion channels [51]. Efforts to enhance delivery by modifying the surface of nanoparticles have shown that uptake can be enhanced by achieving an overall positive zeta potential (+22.1 ± 3.2 mV), versus unmodified nanoparticle negative zeta potential (-27.8 ± 0.5 mV) [47]. However, Nanoparticle uptake under flow conditions has been poorly investigated. In one study, Lin and colleagues [52] used polystyrene nanoparticle coated with a platelet glycoprotein Ibα (GP Ib α– a ligand for P-selectin) to model platelet interaction with endothelial cells and used a flow chamber to examine uptake under flow conditions, using confocal microscopy [52]. Maximum uptake was achieved for the smallest nanoparticle (100nm), and inversely correlated to shear stress levels. Significantly higher cellular uptake was observed for activated human aortic endothelial cells for P-selectin coated surfaces and activated endothelial cells under physiological flow conditions (at 1-5 dyn/cm$^2$ over a 30 minute flow period). In another study, magnetic nanoparticle (magnetite) uptake under flow was examined within isolated bovine femoral arteries [43]. Using an *in vitro* system, they demonstrated that under a magnetic field of 10T/m and at flow rate of 6 ml/min, functionalised magnetite nanoparticles are taken up into endothelial cells [43].

## 3. EFFECTS OF NANOPARTICLE UPTAKE ON VASCULAR FUNCTION AND CONTRACTILITY

### 3.1 Vascular Function and Contractility

A major function of the vasculature is adequate tissue perfusion, to ensure that supply of oxygen and nutrients is matched to tissue demand. This is achieved through the control of vessel diameter, *via* a number of regulatory mechanisms (eg. neural, metabolic, autonomic) that control smooth muscle cell contraction and relaxation. In addition, the endothelial layer lining vessels has an important modulatory function in vessel diameter regulation, through the release of a number of vasodilator and vasoconstrictor substances. An important vasodilator molecule is nitric oxide, which is endogenously produced by endothelial cells through the activation of the endothelial derived nitric oxide synthase (eNOS) enzyme. Nitric oxide is also an inhibitor of smooth muscle cell proliferation and platelet adhesion and aggregation, which are two processes implicated in atherogenesis and restenosis. Other vasodilator factors also exist which lead to vasodilation and contribute to the overall regulation of arterial diameter. These include prostaglandins and a number of endothelial derived hyperpolarising factors (EDHFs) [53, 54]. A number of agonists, including acetylcholine (Ach) can act *via* surface endothelial receptors to stimulate the release of these endothelial derived mediators. Hence, endothelial cells play an essential role in maintaining vascular homeostasis and function through the elaboration of a number of mediators. However, whether uptake of nanoparticles into endothelial cells affects function and mediator release has so far been poorly investigated. As nanoparticles penetrate further into the medial layer of the vasculature, their effects on the contractile function of vascular smooth muscle cells is also unclear. A number of pharmacological agents are available which can be used as tools to examine the integrity and sensitivity of the smooth muscle cell layer. One example is the use of the nitric oxide donor sodium nitroprusside (SNP). This molecule is made up of 5 cyanide groups around an iron core and one nitric oxide group. Uptake of SNP into the smooth muscle cells leads to breakage of the nitric oxide group bonding and its release into the tissue, leading to smooth muscle cell relaxation and consequent vessel dilation [55].

### 3.2 Responses in Aortic Vessels

Rosas-Hernandez and colleagues [25] examined the influence of silver nanoparticles on isolated aortic rings (preconstricted with phenyepherine). They demonstrated that the nanoparticles' influence on aortic vessel function was concentration dependent. At low concentration (5 μg/ml) nanoparticles induced vasoconstriction which blocked any endothelial dependent (Ach) dilator responses, but not endothelial independent (SNP) dilator responses. Conversely, at high concentration (100 μg/ml) nanoparticle lead to a pronounced dilation that was abolished in endothelial denuded vessels and partially inhibited in vessels pre-incubated in L-NAME (an inhibitor of eNOS). However, examination of the prepared nanoparticles under EM showed that the nanoparticles had a wide size range, were irregularly shaped, and had a tendency to aggregate [25]. This makes it difficult to draw conclusions since nanoparticle size heterogeneity may be a factor influencing the observed responses.

We have shown that nanoparticles of different material composition can have different effects on vasodilator responses in rat aortic ring preparations. Uptake of 100 nm silica nanoparticles into endothelial cells, lead to attenuated endothelial-dependent (Ach), but not endothelial-independent (SNP) responses. This contrasts with attenuated SNP responses observed after incubation in gold nanoparticles (12 ± 3 nm). Uptake of silica nanoparticles into endothelial cells was seen to be time and charge dependent (Unpublished observations).

### 3.3 Responses in Small Arteries and the Microvasculature

The uptake of nanoparticles into endothelial cells from different vascular beds is likely to influence vasodilator response differently, since the elaboration of endothelial derived mediators has been shown to be dependent on vessel type, as well as the nature and duration of the stimulus. In terms of vessel type, vasodilator responses of arteries from different species, as well as the size of the vessel, have been shown to involve different mediators [56]. Of particular relevance are small size vessels (100-150 μm active diameter). These are also termed 'resistance' vessels because they have been shown to control over 50% of blood flow [57] into tissues. These arteries exhibit autoregulatory responses to mechanical stimuli, such as pressure and shear stress (due to intralumenal flow). This includes a pressure sensitive intrinsic response to elevations in intravascular pressure (myogenic responsiveness). Although myogenic responsiveness/ tone is largely a property of smooth muscle cell function, endothelial derived mediators are important in the modulation of this response, *via* release of vasodilators. Another important autoregulatory mechanism is flow-induced dilation (FID), which involves the elaboration of a number of endothelial derived mediators, including nitric oxide and EDHF [58].

Studies by Nurkiewicz and colleagues indicate that pulmonary exposure of animals to titanium dioxide nanoparticles (also termed ultrafine particulate matter) led to impaired endothelial dependent dilation of systemic arterioles (skeletal muscle) in a dose dependent manner. Changes in arterial diameter of skeletal arterioles (approx. 40 micrometer diameter) were assessed by intravital microscopy [59]. Free radical scavenging, partially restored microvascular function and nitric oxide production, suggesting that titanium dioxide nanoparticle exposure may lead to reduced nitric oxide bioavailability [60]. A similar attenuation of Ach-dependent vasodilation (but not SNP) was observed in subepicardial arterioles (approx. 150 μm diameter) subsequent to inhalation. Additionally, there was an increase in myogenic tone and impaired endothelial depended FID [61]. Endothelial-dependent dilator responses were restored after incubation with ROS scavengers [62]. These finding suggest that titanium dioxide nanoparticle inhalation leads to their translocation to other tissues including skeletal muscle, the myocardium, and the vasculature, thus affecting their dilator function. Using isolated isobarically mounted mesenteric arteries *in vitro*, we show that direct intravascular, but not extravascular exposure to silica (100 nm) nanoparticle over a thirty minute period, led to their uptake by endothelial cells and concomitant reduction in endothelial dependent ach dilator responses – Fig. **2** (unpublished observations).

To date, there is very limited understanding of the mechanisms whereby nanoparticles influence vascular function. However, pollutant inhalation studies have provided some insight. In the case of veins, pollutant exposure (Diesel exhaust) has been shown to lead to enhanced vasoconstriction and endothelial dysfunction, through uncoupling of eNOS [63]. There is no evidence however that this also takes place in the case of ultrafine particle size exposure. Also there is no evidence to date that confirms such mechanism in arteries and arterioles. Cell culture studies, however, do indicate that this is possible (see section 2.3. for details).

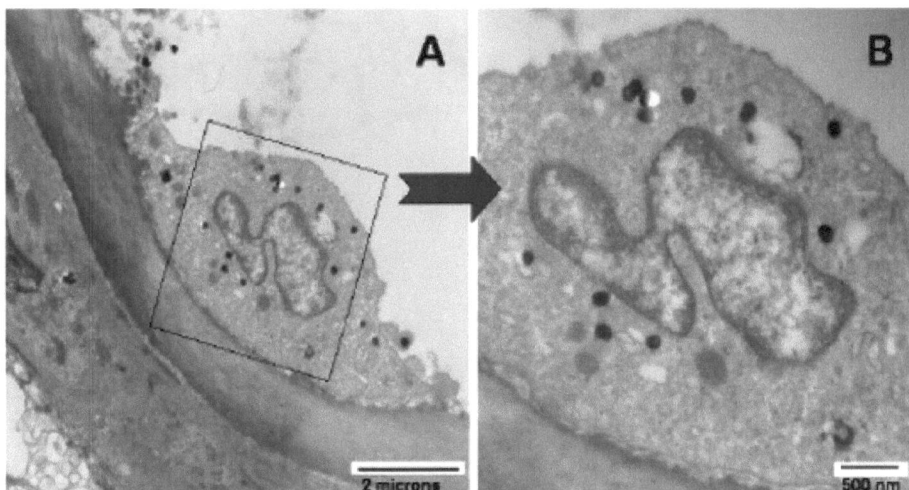

**Figure 2:** Transmission electron micrograph showing silica nanoparticle uptake (100 nm) by endothelial cells of a mesenteric artery after a 30 minute incubation period (A; magnification x10,500); A single endothelial cell electron micrograph at higher magnification, illustrating nanoparticle localisation (dark spheres) in the cytoplasm (B; magnification x25,000).

Another important consideration in assessing the effects of nanoparticles on the vasculature is that effects on the normal healthy vasculature may be different to that for diseased vasculature (which is often targets for therapeutic intervention). Impaired arterial function is evident in a number of disease conditions, including cardiovascular disease, aging, hypertension and diabetes [64]. A key feature of impaired arterial function is the alteration in the balance of endothelial derived vasodilators and vasoconstrictors, leading to 'endothelial dysfunction' and impaired modulation of the function of the underlying smooth muscle cell layer and ultimately control of vessel diameter. In disease states the interplay between the vasodilator pathways may be altered, as in diabetes, where reduced NO mediator release is compensated by upregulation of cyclooxygenase pathways, leading to prostaglandin-induced dilation in coronary arteries from diabetic patients [65]. This highlights the need to examine the influence of nanoparticle uptake on the various signalling pathways in healthy as well as diseased vessels.

## CONCLUSIONS

Previous studies examining the effects of nanoparticle uptake on cellular function, especially cells of the vasculature, have often lacked the availability of nanoparticles with precisely defined size, charge and characteristics. Consequently, reports by both DEFRA (2007) and the international life sciences institutes (ILSI) research foundation have highlighted the need to consider the physicochemical characteristics of nanoparticles as an essential component in toxicity testing [11, 66]. More recently, the increased availability of characterisation tools with higher spacial resolution, such as atomic force microscopy, Raman spectroscopy, transmission and scanning electron microscopy, and the use of Zeta sizers for charge and zeta potential measurements, now allows for precise determination of the physicochemical properties of functionalised nanoparticles. This is necessary for accurate interpretation of findings, especially in relation to the surface properties of nanoparticles which can vary according to the medium in which the nanoparticles are suspended. Other recommendations still require addressing, which include the need to carry out detailed testing in cell-free, *in vitro* cell culture as well as *in vivo* conditions [67], using nanoparticle concentrations that match normal exposure levels [1]. Studies in both the normal and compromised/ diseased state are necessary, since nanoparticle penetration properties into tissues can vary accordingly [1]. Detailed toxicity studies will allow us to identify those nanoparticle materials, concentrations and sizes which will induce least cellular and tissue toxicity. In order to adequately assess the efficiency of therapeutic intervention it is necessary to determine the nanoparticle properties that are likely to have minimal detrimental effects on vascular function and contractility, to help maintain tissue perfusion and healthy organ function.

## ACKNOWLEDGEMENTS

The author wishes to thank Dr. Debra Whitehead (Lecturer in the division of Chemistry, School of Biology, Chemistry and Health Science, Manchester Metropolitan University) for the supply and characterisation of the silica and gold nanoparticles; Dr. Carolyn Jones (Maternal and Fetal Health Research Group, University of Manchester) for electron microscopy; and Mo Alkattan for illustrations.

## REFERENCES

[1]    Oberdorster G. Safety assessment for nanotechnology and nanomedicine: concepts of nanotoxicology. J Int Med 2009;267:89-105.

[2].   Kim YI, Fluckiger L, Hoffman M, *et al.* The antihypertensive effect of orally administered nifedipine-loaded nanoparticles in spontaneously hypertensive rats. Br J Pharmacol 1997;120(3):399-404.

[3]    Reddy GR, Bhojani MS, McConville P, *et al.* Vascular targeted nanoparticles for imaging and treatment of brain tumors. Clin Cancer Res 2006;12(22):6677-86.

[4]    Reddy MK, Labhasetwar V. Nanoparticle-mediated delivery of superoxide dismutase to the brain: an effective strategy to reduce ischemia-reperfusion injury. FASEB J 2009;23(5):1384-95.

[5]    Nishikawa T, Iwakiri N, Kaneko Y, *et al.* Nitric oxide release in human aortic endothelial cells mediated by delivery of amphiphilic polysiloxane nanoparticles to caveolae. Biomacromolecules 2009;10(8):2074-85.

[6]    Chen J, Patil S, Seal S, *et al.* Rare earth nanoparticles prevent retinal degeneration induced by intracellular peroxides. Nat Nanotechnol 2006;1(2):142-50.

[7]    Margolis J, McDonald J, Heuser R, *et al.* Systemic nanoparticle paclitaxel (nab-paclitaxel) for in-stent restenosis I (SNAPIST-I): a first-in-human safety and dose-finding study. Clin Cardiol 2007;30(4):165-70.

[8]    Cho M, Cho WS, Choi M, *et al.* The impact of size on tissue distribution and elimination by single intravenous injection of silica nanoparticles. Toxicol Lett 2009;189(3):177-83.

[9]    De Jong WH, Hagens WI, Krystek P, *et al.* Particle size-dependent organ distribution of gold nanoparticles after intravenous administration. Biomaterials 2008 ;29(12):1912-9.

[10]   Sonavane G, Tomoda K, Makino K. Biodistribution of colloidal gold nanoparticles after intravenous administration: effect of particle size. Colloids Surf B Biointerfaces 2008;66(2):274-80.

[11]   DEFRA, 2007. Characterising the potential risks posed by engineered nanoparticles. A second UK government research report. Available at: www.defra.gov.uk/environment/nanotech/research/pdf/nanoparticles-riskreport07.pdf

[12]   Royal Society and Royal Academy of Engineering, UK, 2004. Nanoscience and nanotechnologies: opportunities and uncertainties. Available at http://www.roysoc.ac.uk

[13]   Suh H, Jeong B, Liu F, Kim SW. Cellular uptake study of biodegradable nanoparticles in vascular smooth muscle cells. Pharm Res 1998;15(9):1495-8.

[14]   Panyam J, Labhasetwar V. Dynamics of endocytosis and exocytosis of poly(D,L-lactide-co-glycolide) nanoparticles in vascular smooth muscle cells. Pharm Res 2003;20(2):212-20.

[15]   Davda J, Labhasetwar V. Characterization of nanoparticle uptake by endothelial cells. Int J Pharm 2002;233(1-2):51-9.

[16]   Panyam J, Zhou WZ, Prabha S, *et al.* Rapid endo-lysosomal escape of poly(DL-lactide-co-glycolide) nanoparticles: implications for drug and gene delivery. FASEB J 2002;16(10):1217-26.

[17]   Rejman J, Oberle V, Zuhorn IS, Hoekstra D. Size-dependent internalization of particles *via* the pathways of clathrin- and caveolae-mediated endocytosis. Biochem J 2004;377(1):159-69.

[18]   Chithrani BD, Ghazani AA, Chan WC. Determining the size and shape dependence of gold nanoparticle uptake into mammalian cells. Nanoletters 2006;6:662-668.

[19]   Zhang S, Li J, Lykotrafitis G, *et al.* Size-dependent endocytosis of nanoparticles. Adv Mater Deerfield 2009;21:419-424.

[20]   Chithrani DB, Dunne M, Stewart J, *et al.* Cellular uptake and transport of gold nanoparticles incorporated in a liposomal carrier. Nanomedicine 2010;6(1):161-9.

[21]   Geiser M, Rothen-Rutishauser B, Kapp N, *et al.* Ultrafine particles cross cellular membranes by nonphagocytic mechanisms in lungs and in cultured cells. Environ Health Perspect 2005;113(11):1555-60.

[22]   Wang Z, Tiruppathi C, Minshall RD, Malik AB. Size and dynamics of caveolae studied using nanoparticles in living endothelial cells. ACS Nano 2009;3(12):4110-6.

[23]   Fröhlich E, *et al.* Cytotoxicity of nanoparticles independent from oxidative stress. J Toxicol Sci 2009;34(4):363-75.

[24]   Oesterling E, *et al.* Alumina nanoparticles induce expression of endothelial cell adhesion molecules. Toxicol Lett 2008;178(3):160-6.

[25]   Rosas-Hernández H, Jiménez-Badillo S, Martínez-Cuevas PP, *et al.* Effects of 45-nm silver nanoparticles on coronary endothelial cells and isolated rat aortic rings. Toxicol Lett 2009;191(2-3):305-13.

[26]   Park EJ, Park K. Oxidative stress and pro-inflammatory responses induced by silica nanoparticles *in vivo* and *in vitro*.Toxicol Lett 2009;184(1):18-25.

[27]   Cho WS, Cho M, Jeong J *et al.* Acute toxicity and pharmacokinetics of 13 nm-sized PEG-coated gold nanoparticles. Toxicology and App Pharmacol 2009;236:16-24.

[28]   Hudson SP, Padera RF, Langer R, *et al.* The biocompatibility of mesoporous silicates. Biomaterials 2008;29(30):4045-55.

[29]   Panyam J, Sahoo SK, Prabha S, *et al.* Fluorescence and electron microscopy probes for cellular and tissue uptake of poly(D,L-lactide-co-glycolide) nanoparticles. Int J Pharm 2003;262(1-2):1-11.

[30].  Jin Y, Kannan S, Wu M, *et al.* Toxicity of luminescent silica nanoparticles to living cells. Chem Res Toxicol 2007;20(8):1126-33.

[31]   Slowing II, Wu CW, Vivero-Escoto JL, *et al.* Mesoporous silica nanoparticles for reducing hemolytic activity towards mammalian red blood cells. Small 2009;5(1):57-62.

[32]   Huang X, Teng X, Chen D, *et al.* The effect of the shape of mesoporous silica nanoparticles on cellular uptake and cell function. Biomaterials 2010;31(3):438-48.

[33]   Heikkilä T, Santos HA, Kumar N, *et al.* Cytotoxicity study of ordered mesoporous silica MCM-41 and SBA-15 microparticles on Caco-2 cells. Eur J Pharm Biopharm 2010;74(3):483-94.

[34]   Gojova A, Guo B, Kota RS, *et al.* Induction of inflammation in vascular endothelial cells by metal oxide nanoparticles: effect of particle composition. Environ Health Perspect 2007;115(3):403-9.

[35]   Kennedy IM, Wilson D, Barakat AI; HEI Health Review Committee. Uptake and inflammatory effects of nanoparticles in a human vascular endothelial cell line.Res Rep Health Eff Inst. 2009 ;(136):3-32.

[36]   Connor EE, Mwamuka J, Gole A, *et al.* Gold np are taken up by human cells but do not cause acute cytotoxicity. Small 2005;1:325-327.

[37]   Kim JH, Kim JH, Kim KW, *et al.* Intravenously administered gold nanoparticles pass through the blood-retinal barrier depending on the particle size, and induce no retinal toxicity. Nanotechnology. 2009 ;20(50):505101.

[38]   Stelzer R, Hutz RJ. Gold nanoparticles enter rat ovarian granulosa cells and subcellular organelles, and alter in-vitro estrogen accumulation. J Reprod Dev 2009;55(6):685-90.

[39]   Mironava T, Hadiargyrou M, Simon M, *et al.* Gold nanoparticles cellular toxicity and recovery: Effect of size, concentration and exposure time. Nanotoxicity 2010;4(1):120-137.

[40]   Hung HS, Wu CC, Chien S, *et al.* The behavior of endothelial cells on polyurethane nanocomposites and the associated signaling pathways.Biomaterials 2009;30(8):1502-11.

[41]   Li R, Ning Z, Cui J, *et al.* Ultrafine particles from diesel engines induce vascular oxidative stress *via* JNK activation. Free Radic Biol Med 2009;46(6):775-82.

[42]   Gojova A, Lee JT, Jung HS, *et al.* Effect of cerium oxide nanoparticles on inflammation in vascular endothelial cells. Inhal Toxicol. 2009 ;21 Suppl 1:123-30.

[43]   Seliger C, Jurgons R, Wiekhorst F, *et al. In vitro* investigation of the behaviour of magnetic particles by a circulating artery model. J Magn Magnetic Materials 2007;311:358-362.

[44]   Jani P, Halbert GW, Langridge J, *et al.* Nanoparticle uptake by the rat gastrointestinal mucosa: quantitation and particle size dependency. J Pharm Pharmacol 1990;42(12):821-6.

[45]   Nemmar A, Hoylaerts MF, Hoet PH, *et al.* Possible mechanisms of the cardiovascular effects of inhaled particles: systemic translocation and prothrombotic effects. Toxicol Lett 2004;149(1-3):243-53.

[46]   Manolova V, Flace A, Bauer M, *et al.* Nanoparticles target distinct dendritic cell populations according to their size. Eur J Immunol 2008;38(5):1404-13.

[47]   Labhasetwar V, Song C, Humphrey W, *et al.* Arterial uptake of biodegradable nanoparticles: effect of surface modifications.J Pharm Sci. 1998 ;87(10):1229-34.

[48]   Oberdörster G, Oberdörster E, Oberdörster J. Nanotoxicology: an emerging discipline evolving from studies of ultrafine particles. Environ Health Perspect 2005;113(7):823-39.

[49]   Cai QY, Cai QY, Kim SH, *et al.* Colloidal gold nanoparticles as a blood-pool contrast agent for X-ray computed tomography in mice. Invest Radiol 2007;42:797-806.

[50]   Ma HL, *et al.* Superparamagnetic iron oxide np stabilized by alginate: pharmacolkinetics, tissue distribution and applications in detecting liver cancers. Int J Pharm 2008;16:217-226.

[51]   Westedt U, Barbu-Tudoran L, Schaper AK, *et al.* Deposition of nanoparticles in the arterial vessel by porous balloon catheters: localization by confocal laser scanning microscopy and transmission electron microscopy. AAPS PharmSci 2002;4(4):E41.

[52]   Lin A, Sabnis A, Kona S, *et al.* Shear-regulated uptake of nanoparticles by endothelial cells and development of endothelial-targeting nanoparticles. J Biomed Mater Res A. 2010 ;93(3):833-42.

[53]   Matsumoto T, Miyamori K, Kobayashi T, *et al.* Specific impairment of endothelium-derived hyperpolarizing factor-type relaxation in mesenteric arteries from streptozotocin-induced diabetic mice. Vascular Pharmacol 2006;44(6):450-60.

[54]   Pannirselvam M, Ding H, Anderson TJ, *et al.* Pharmacological characteristics of endothelium-derived hyperpolarizing factor-mediated relaxation of small mesenteric arteries from db/db mice. Eur J Pharmac 2006;551(1-3):98-107.

[55]   Smith JN, Dasgupta TP. Mechanisms of nitric oxide release from nitrovasodilators in aqueous solution: reaction of the nitroprusside ion ([Fe(CN)5NO]2-) with L-ascorbic acid. J Inorg Biochem 2001;87(3):165-73.

[56].   Ingram DG, Newcomer SC, Price EM, *et al.* Chronic nitric oxide synthase inhibition blunts endothelium-dependent function of conduit coronary arteries, not arterioles. Am J Physiol Heart Circ Physiol 2007;292(6):H2798-808.

[57]   Chilian WM, Layne SM, Klausner EC, *et al.* Redistribution of coronary microvascular resistance produced by dipyridamole. Am J Phys 1989;256:H383-90.

[58]   Azzawi M, Austin C. Transient vasodilatory responses to intralumenal flow in isolated pressurised rat coronary arteries: influences of different endothelial factors. J Vasc Res 2007;44:223-233.

[59]   Nurkiewicz TR, Porter DW, Hubbs AF, *et al.* Nanoparticle inhalation augments particle-dependent systemic microvascular dysfunction. Part Fibre Toxicol 2008;5:1.

[60]   Nurkiewicz TR, Porter DW, Hubbs AF, *et al.* Pulmonary nanoparticle exposure disrupts systemic microvascular nitric oxide signaling. Toxicol Sci 2009;110(1):191-203.

[61]   LeBlanc AJ, Cumpston JL, Chen BT, *et al.* Nanoparticle inhalation impairs endothelium-dependent vasodilation in subepicardial arterioles. J Toxicol Environ Health A 2009;72(24):1576-84

[62]   LeBlanc AJ, Moseley AM, Chen BT, *et al.* Nanoparticle inhalation impairs coronary microvascular reactivity *via* a local reactive oxygen species-dependent mechanism. Cardiovasc Toxicol 2010;10(1):27-36.

[63]   Knuckles TL, Lund AK, Lucas SN, *et al.* Diesel exhaust exposure enhances venoconstriction *via* uncoupling of eNOS. Toxicol Appl Pharmacol 2008;230(3):346-51.

[64]   Fang ZY, Prins JB, Marwick TH. Diabetic cardiomyopathy: Evidenc, Mechanism, and therapeutic implications. Endocrine Rev 2004;25:543-567.

[65]   Szerafin T, Erdei N, Fülöp T, *et al.* Increased cyclooxygenase-2 expression and prostaglandin-mediated dilation in coronary arterioles of patients with diabetes mellitus. Circ Res 2006;99:e12-7.

[66]   Oberdorster G, MaynardA, Donaldson K *et al. and a report from the ILSI Research Foundation/Risk Science Institute Nanomaterial Toxicity Screening Working group.* Principles for characterising the potential human health effects from exposure to nanomaterials: elements of a screening strategy. Part Fibre Toxicol 2005;2:8.

[67]   Fischer HC, Chan WCW. Nanotoxicity: the growing need for *in vivo* study. Curr Opin Biotech 2007;18:565-571.

# CHAPTER 7

# Nanorobots for Endovascular Target Interventions in Future Medical Practice

## Sylvain Martel*

*École Polytechnique de Montréal (EPM), Campus of the University of Montréal, Montréal (Québec), Canada*

**Abstract:** Medical robotics have evolved from interventions being performed by relatively large robots outside the patient to smaller untethered versions such as the camera pills for operations inside the digestive track. But despite these recent technological advances, many types of medical interventions are still out-of-reach to such modern medical robotic systems. More recently, a new class of untethered robots has emerged. These robots are miniaturized further to target regions in the human body only accessible through smaller diameter blood vessels and as such, they could play a more critical role in many medical applications. Tumor targeting is an obvious example where robotics being applied to miniature untethered carriers capable of transporting drugs can play a major role by offering an improved concentration of therapeutic agents at the targeted area and a decrease of systemic side effects compared to modern interventions such as chemotherapy. But to be successful, such class of robotics must as nanomedicine did, consider nanotechnology to implement critical functionalities aimed at enabling new target therapies or at least to improve many existing medical interventions. This new field of robotics referred here to as medical nanorobotics would therefore achieve such goal by embedding and exploiting nanometer-scale components and phenomena within the context of robotics. In this chapter, the nanoparticles being among the simplest nanometer-scale components to be embedded in such miniature robots will be used as an example to initiate the readers with the powerful concept of nanorobotics and how it can affect future medical practice and in particular, endovascular target interventions.

**Keywords:** Nanorobot; camera; drug transportation; endovascular; magnetic nanoparticles; drug delivery and targeting; tumour delivery; functional; magnetic fields.

## 1. MEDICAL ROBOTICS AT DIFFERENT SCALES

Because of continuous advances in science and technology, robotics is still evolving, offering new capabilities and potential in a more diverse range of applications including medical interventions. The potential for robotics in medical practice has been first recognized with the use of surgical robots (see Fig. **1a**) such as the well known da Vinci platform [1]. This type of robot was followed later by a smaller device known as the camera pill [2] (see Fig. **1b**) designed to travel in the human digestive track after being swallowed by the patient. Although most people would not argue that the first example can be considered as a robot, many would question if the camera pill can in fact, be defined as a robot. The Robotics Institute of America (RIA) for instance gives a similar but slightly broader version of the definition of a robot provided by the International Organization of Standardization in the ISO 8373 which has been used by the International Federation of Robotics, the European Robotics Research Network (EURON), and many national standards committees. It defines a robot as a re-programmable multi-functional manipulator designed to move materials, parts, tools, or specialized devices through variable programmed motions for the performance of a variety of tasks. Accordingly, some versions so the camera pills and especially the more recent ones that have embedded functionalities could be defined as robots.

But as depicted in Fig. **1**, scaling has a huge impact on the design of a robot which leads to different approaches in their implementations. One important factor among others that influence the design at the

---

*Address correspondence to Sylvain Martel: École Polytechnique de Montréal (EPM), Campus of the University of Montréal, P.O. Box 6079, Station Centre-Ville, Montréal, Qc., Canada, H3C 3A7; Tel: (514) 340-4711 ext.: 5098; E-mail: sylvain.martel@polymtl.ca

micrometer-scale and beyond is that powering medical miniature robots remains a huge issue and although research is currently focusing on how devices gather and store energy, no practical solution is still available. As a result, much smaller implementations deviate substantially in many respects from the larger scale versions. While larger robots are traditionally based on mechanical and electronic components, micrometer-scale and smaller implementations rely more on material properties, biology, and nano-components, and despite that they do not appear as traditional robots, in many instances, they may be considered as robots. For example, Fig. **1c** depicts a version of an untethered entity (only a few tens of micrometers across) that fulfills the above definition of a robot. Indeed, this particular example based on nanoparticles embedded in aggregated microscale polymeric particles is capable (as described in more details later) of programmable motion for the delivery of therapeutic agents at a specific location in the human body. For the latter example, if such aggregated microparticles would be simply injected in the body without programmable motion and/or functions, then they would not be defined as robots. As such, when such untethered entities are designed to be interfaced with an infrastructure or platform allowing programmable functions/motion, then such entities can be considered as being integral components of a robotic platform. Going further, a 2-micrometer (μm) in diameter flagellated MC-1 bacterium [3] as depicted in Fig. **1d** and acting under the influence of an external programmable controller to direct its motion for the delivery of therapeutic agents as described in [4] can also be defined as a microscale robot, bio-robot, and/or bacterial robot, which would not be true if the same bacteria could not be controlled, here by a software program executed by an external computer. But although the previous examples can be defined as robots, they cannot be viewed as fully-autonomous robots. Indeed, most people think of miniature robots in the body as fully autonomous small entities that swim in the blood streams without interactions with the external world. This futuristic vision is presently beyond technological limitations and contributes to a misconception of modern medical nanorobotics. Today, it may be possible to have miniature entities with similar characteristics and functionalities as some of these futuristic versions, but only if such entities rely on an external infrastructure. Although fully-autonomous implementations of such miniature robots capable of travelling in the human microvasculature are beyond present technological limits, present implementations can be quite powerful and make a significant impact to the medical practice, especially when nano-components are being considered. Indeed, even at the smallest scale and as explained later in this chapter, nanoparticles (Fig. **1e**) can enhance the medical impact of such miniature robots, allowing several programmable functions to be implemented.

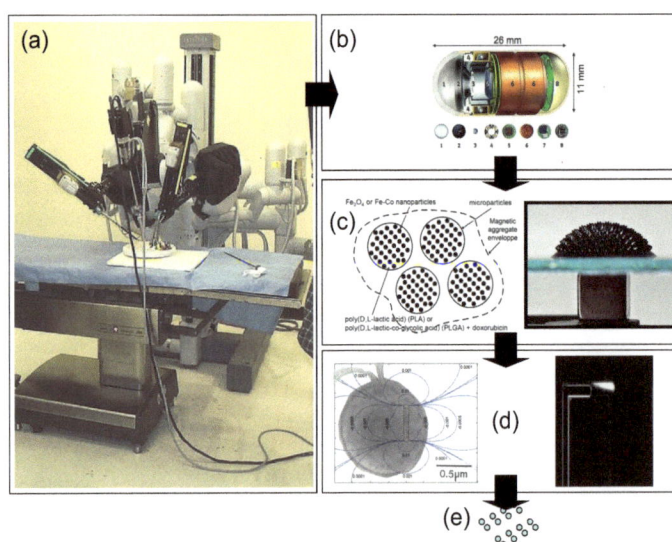

**Figure 1:** (a) A photograph of a surgical robotic platform (from [5]); (b) The Camera Pill (from [6]); (c) An example of a microscale robot capable of programmable motion in the vascular network with a photograph showing the influence on a large aggregate submitted to the influence of a magnetic field; (d) An electron image of a MC-1 magnetotactic bacterium capable of programmable directional motion provided by a external computer with an image of an aggregate of these bacteria (white spot) acting under computer directional motion control and observed under an optical microscope; and (e) Smallest versions based on aggregates of nanoparticles.

## 2. NANOROBOTICS: NANOTECHNOLOGY AND ROBOTICS

According to the National Nanotechnology Initiative (NNI), nanotechnology is described as the understanding and control of matter at dimensions between approximately 1 and 100 nanometers (nm), where unique phenomena enable novel applications. Encompassing nanoscale science, engineering, and technology, nanotechnology involves imaging, measuring, modeling, and manipulating matter at this length scale. As such, the exploitation of nanotechnology can become a powerful concept in the field of robotics since physical, chemical, and biological properties can emerge in materials at the nanoscale, and may differ in important ways from the properties of bulk materials and single atoms or molecules. Indeed, because nanoscale materials also have size-dependent magnetic behavior, mechanical properties, and chemical reactivity, their integration into small scale devices or robots can enable new functionalities that would otherwise be impossible to implement. A simple example of such size-dependent behaviors is magnetic nanoparticles (MNP). At a scale of a few nanometers, magnetic nanoparticles (or nanoclusters) have a single magnetic domain, and the strongly coupled magnetic spins on each atom combine to produce a particle with a "giant" spin. For example, the giant spin of a ferromagnetic iron particle rotates freely at room temperature for diameters below approximately 16nm, an effect termed superparamagnetism. It is then obvious that nanoscale object offer a new set of tools for engineers because of the different behaviors compared to macroscopic counterparts. Indeed, at the nanoscale, atomic forces and chemical bonds dominate while macroscopic effects such as convection, turbulence, and momentum (inertial forces) become negligible.

As such, a robot that depends on nanometer-scale components (typically less than 100nm across) to embed characteristics that will allow such a robot to accomplish specific functions that would be impossible to implement otherwise, could be defined as a nanorobot. With the integration of many of these nanometer-scale components, such nanorobot could increase in volume to reach the micrometer-scale. The latter could be defined as a microscale nanorobot unlike a nanoscale nanorobot that would typically have limited embedded functionalities and/or capabilities compared to its microscale counterpart. Therefore, the use of nanoscale components instead of larger components is critical for many functions that need to be embedded in such microscale nanorobots. For example and as described in more details later in this chapter, aggregated MNP can as a single microscale magnetic particle, allow propulsion of untethered robots in the vascular network but only MNP of a specific size range can be used to implement many other functions such as hyperthermic-based actuation in the human body. In turn, hyperthermic-based actuations can be used for many purposes, such as for the implementation of polymorphic robots, to computer-controlled embolization and drug release coupled with enhanced therapeutic efficacy achieved by controlled local increase in temperature.

Dimensions in the nanometers also matter when specific targets in the human body must be reached. In other words, the environment where the operation is performed may also influence the size of the robotic components being considered. For instance, studies indicate that nanoparticles in the 7-10 nm size range are smaller than the 12 nm physiological upper limit of pore size within the blood-tumor barrier which allows an accumulation of high concentrations within individual cancer cells. Adjusting the robotic platform to perform hyperthermia on MNP of equivalent size is highly feasible and will increase the range of functionalities being supported, providing more opportunities for enhanced therapeutic efficacy while at the same time, influencing future medical practice by defining new interventional protocols adapted to such technology.

But nanotechnology can not only be used to implement functionalities in such nanorobots but the latter can also benefice from progress in nanomedicine. Simply defined, nanomedicine is nanotechnology applied to medicine. Nanomedicine can be further defined as the monitoring, repair, construction and control of human biological systems at the molecular level, using engineered nanodevices and nanostructures. The encapsulation or attachment of specific proteins, antibodies or RNA strands, are only a few examples that may enhance further the treatment efficacy when interfaced to nanorobots involved in target interventions.

## 3. ROBOTS INSIDE THE BODY

Several regions inside the human body could theoretically be accessed by miniature untethered robots. Presently, intensive efforts in research and development are being done for operations in specific regions inside the human body. This is depicted in Fig. **2**. Going back to one of our previous examples, the camera pill is the first commercially available system that can be regards as an untethered robot capable of operating inside the human body. The camera pill is primarily used as a miniature untethered endoscope to inspect the human gastrointestinal (GI) track. Several versions have been produced and others aimed at increasing the performance and/or the level of functionality are still in the research phase. The VECTOR pill [7] with three versions shown in Fig. **2** and designed for advanced diagnostics and therapy in the human digestive tract is one example. Besides the GI track, interventions inside the human eye have been investigated [8]. Indeed, external magnetic fields have been used to steer one-millimeter-long robots inside the human eye for sensing, drug delivery and surgery. Using a microrobot could modify the medical practice for eye surgeons. The delivery of drugs directly to veins inside the retina is one example of potential applications. As in most implementations at such a scale, the choice of material being used is important. This microrobot is built from fine layers of indium, gallium, arsenic and chrome, with its "head" made of chrome, nickel and gold. As nickel is slightly magnetic, the robot can be moved in low magnetic fields up to a speed of 20μm/s.

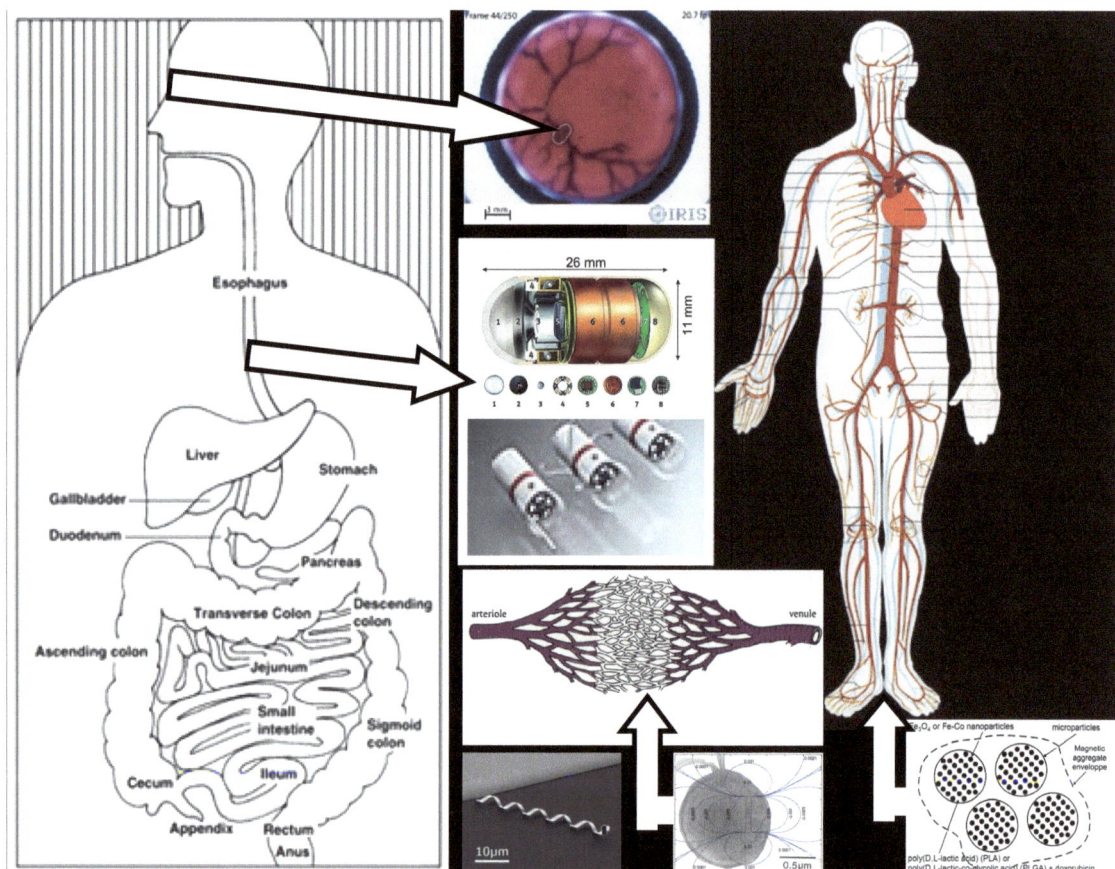

**Figure 2**: (Left) The eye with the respective microrobot (Top-middle) and the GI tract with some of the robots (the camera pill with three versions of an interventional prototype [7]) as the main accessible regions being under investigation; (Right) The larger vessels of the vascular network with an aggregate of TMMC [9]; (Bottom-center) an artificial flagellum [10] (left) and a natural flagellated bacterium [4] (right) considered for the delivery of therapeutic agents through the microvasculature.

But for maximum accessibility inside the human body, the vascular network is certainly, compared to other potential routes, the one that offers the best accessibility to the various regions in the body. Indeed, such network is composed of nearly 100,000km of blood vessels with diameters varying from a few millimeters in the arteries, down to ~4μm in the smaller capillaries found in the microvasculature, with respective important variations in blood flow velocities. Efforts are being done to synthesize microscale entities that can be navigated in such network [4,9,11-12]. In larger diameter vessels, such entities rely on aggregates of microparticles encapsulating MNP as depicted in the lower right corner of Fig. **2**. The embedded MNP are used for MRI-based propulsion/steering (referred to as Magnetic Resonance Propulsion (MRP)), tracking with MRI, and hyperthermic-based actuation, to name but the main functions. When the vessel's diameter decreases substantially when deeper regions in the body are targeted, the diameter of each robot must decrease accordingly and as such, the volume of embedded magnetic material becomes too small to allow for a level of MRP sufficient for targeting purpose. Therefore, propulsion techniques relying on external magnetic field gradients must be replaced in the microvasculature by more efficient systems under such conditions and self-propelled microsystems such as flagella-based propulsions as found in nature appear to be powerful candidates. Indeed, flagellar propulsion as used in many species of bacteria has proven to be more effective in low Reynolds hydrodynamics (this is explained later) as it is the case when operating in the microvasculature. Two approaches are presently investigated. One proposes the use of artificial flagella [10] (see photograph in Fig. **2**) which still rely on an external magnetic field but of less magnitude, while the other relies on self-propelled (*i.e.* with no need for an external source for propulsion) natural flagella [4] and more specifically on the use of flagellated MC-1 magnetotactic bacteria (MTB) being controlled by computer (see photograph in Fig. **2**). This is discussed in more details later in this chapter.

## 4. TARGET DRUG DELIVERY USING CLOSED-LOOP NAVIGATION CONTROL

Some applications of nanomedicine include but are not limited to cell/gene therapy, biomarkers, medical imaging, molecular diagnostics, genomics/proteomics, nano-biopharmaceuticals, drug discovery, and drug delivery. The latter applied to cancer therapy will be used here as an example of applications to show the potential of medical nanorobotics in future medical practice.

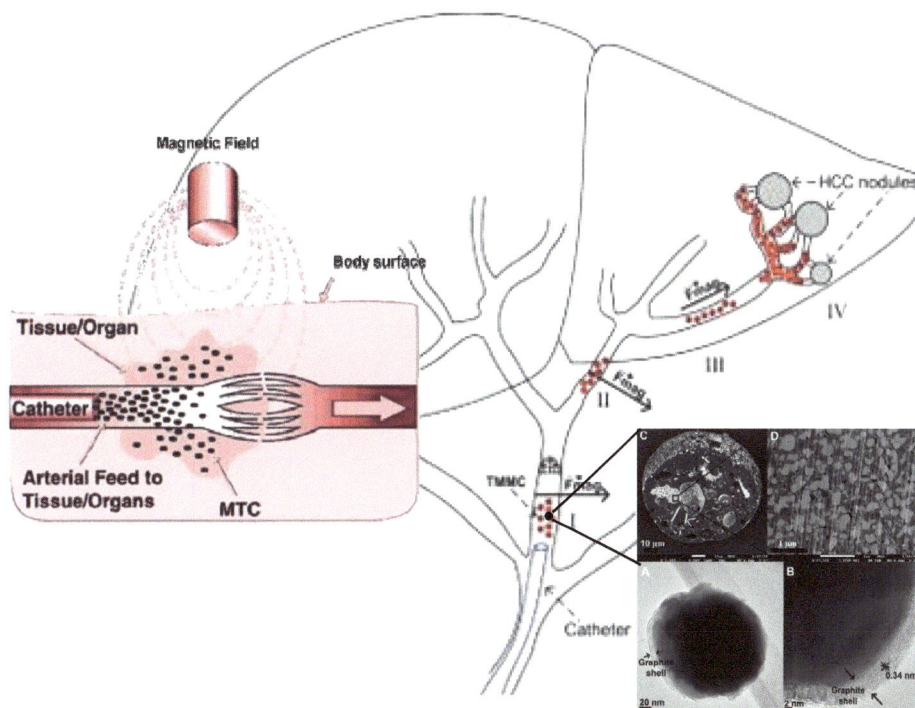

**Figure 3:** (Left) Use of a magnet for targeting tumors (figure from [13]); (Middle) Navigation of TMMC from the catheter release site to the liver tumor; (Bottom-right) photograph of the TMMC with (A) embedded MNP, (B) close-up view of the MNP, (C) the TMMC with (D) a close-up view of the TMMC (images from [12])

The ultimate goal of cancer therapy is to develop agents that will selectively destroy cancer cells, sparing the normal tissues of the patient. Most current available cancer chemotherapeutic agents target DNA or the enzymes involved in DNA replication. As such, they destroy all rapidly dividing cells, including normal dividing cells in vital tissues. Therefore, the low chemotherapeutic index of these agents leads to severe generalized toxic effects when used at dosages necessary to kill tumor cells. Such dosage in the systemic circulation can be considered extremely high and damageable due to a high level of secondary toxicity, not only applies in the cases of cancer being spread in the body (metastasis), but also in the presence of localized solid tumors. In the latter cases, because of the lack of methods for direct targeting, a substantial amount of therapeutics is injected in the systemic circulation to ensure that a sufficient percentage of agents for therapeutic purpose will reach the tumoral lesions. We define direct targeting here, the capability of targeting a specific region using the most appropriate and/or direct route as to avoid as much as possible the systemic circulation and hence, eliminating or at least minimizing secondary toxicity. The latter can be done through closed-loop control, a fundamental requirement in robotics.

Efforts have been done to guide lower dosage of therapeutic agents toward a tumor. One state-of-the-art approach is depicted in Fig. **3** where a permanent magnet is placed near the targeted region and above (outside) the body surface [13]. As depicted on the left of Fig. **3**, a catheter is used to release Magnetic Therapeutic Carriers (MTC). This approach has major limitations and drawbacks especially for targets located deeper below the skin surface. As shown in Fig. **3**, the permanent (or an electro-magnet) located near the tumor and outside the body provides higher field intensity towards the external magnet with much higher intensity near the magnet. Hence, effective targeting becomes mostly restricted to tumors near the skin. When the targeted region is located deeper in the body, a significant reduction of targeting efficacy is likely to occur. Furthermore, targeting efficacy is significantly affected since the approach relies on magnetic trapping toward the skin surface without trajectory servo-control where the displacement of magnetic carriers can be influenced and corrected during travel based on tracking information from the catheterization release site and towards the targeted region. The reachable limits of catheterization due to many factors including the decreasing diameter of the blood vessels, and the number of bifurcations towards the targeted region caused by complex microvasculature networks especially near the tumor, will combine to contribute to lower significantly the efficacy in targeting unless proper closed-loop navigation control is performed. A simple example of such closed-loop targeting is depicted in the right portion of Fig. **3**.

The direction of motion of navigable entities containing therapeutic agents and referred here to as Therapeutic Magnetic Micro Carriers (TMMC) [12] (see Fig. **3**), is influenced by magnetic gradients generated by an upgraded clinical MRI platform [9] capable of sufficient propulsion/steering gradient amplitude within FDA regulations (TMMC being one form of microscale nanorobots). In this simple example, TMMC are guided from the catheterization boundary in an artery to the embolization site where the drug is released. Without closed-loop navigation control, at each bifurcation, an increasing proportion of the therapeutic agents would miss the targeted embolization site and be dispersed, decreasing substantially the therapeutic efficacy while increasing substantially secondary toxicity. As such, closed-loop control can substantially prevent secondary toxicity while improving therapeutic efficacy with a lower dosage by directing TMMC in the appropriate branch at each vessel's bifurcations. This principle has been described and with experimental results performed *in vivo* for the first time using a clinical MRI scanner in [14].

A closed-loop transfer function with a basic schematic of a closed-loop navigation control scheme (also referred as navigation feedback or servo-control) for target interventions in the vascular network are depicted in Fig. **4**. In Fig. **4**, the output of the system is fed back through a sensor measurement to the reference value. In this case, the sensor is the tracking signal generated by the MNP embedded in each microscale nanorobots and captured by a MRI scanner. The controller executes a software control algorithm [15] in a computer being linked to the MRI scanner [16]. It then takes tracking information and compares it to the planned position or trajectory (the reference). Such information can also be displayed to the user (medical staff) as depicted in Fig. **4**. The error (difference) between the reference and the measured output (the tracked position of the microscale nanorobots) is used to change the input to the system (here the aggregate of microscale nanorobots) under control, creating a system output, *i.e.* causing a displacement of the nanorobots to a new position. This closed-loop process must be executed at a rate sufficiently high for effective navigation within the conditions imposed by the regions being navigated within the vascular network. The overall latency (*i.e.* the delay in gathering the required tracking information to the time at which the directional force created by the magnetic gradients is induced on the MNP to influence the motion of the nanorobots) must also be sufficiently short. Since many parameters and variables are taken into account at a relatively high frequency and low latency, it becomes obvious that such real-time closed-loop navigation process must be performed by computer with no (or at least with minimum) human interventions during the real-time process. With such an approach, interventions from medical specialists are expected to occur during the non real-time phases, *i.e.* during the pre-operative (planning) and post-operative phases.

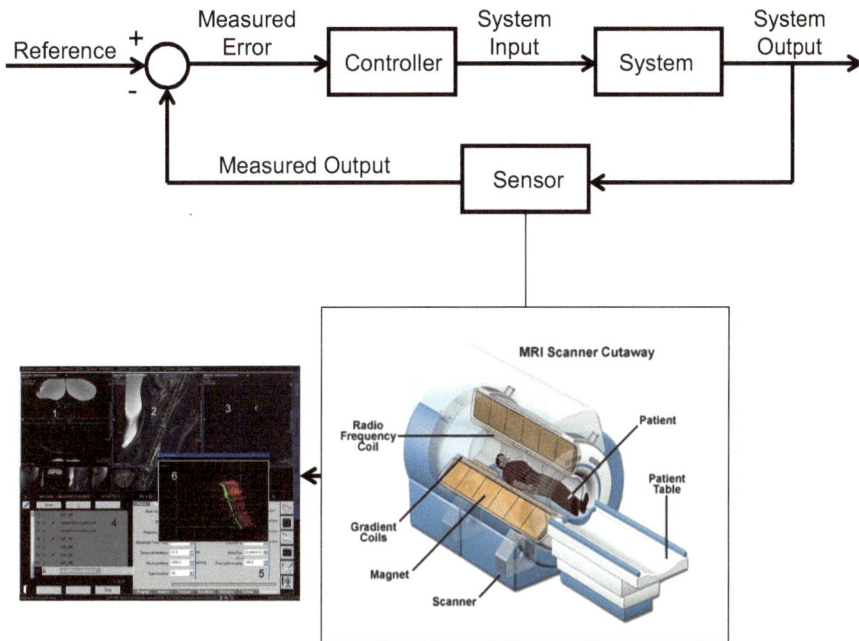

**Figure 4:** (Top) A classical closed-loop control scheme; (Bottom-right) An image of a clinical MRI scanner used as the sensor during closed-loop navigation (image taken from MRI: A Guided Tour by Kristen Coyne); (Bottom-left) one example of a computer user interface during the real-time navigation in an upgraded MRI scanner (from [9])

## 5. MEDICAL IMAGING MODALITIES

Closed-loop navigation control is effective only if tracking information from inside the vascular network to determine the 3D position of the nanorobots can be fed back to the controller. Such information must also be fed back at a speed sufficiently high to allow corrective propulsion/steering force to be applied in a timely manner for targeting purpose. In this respect, traditional robotic tracking methods based on direct line-of-sight as used in other robotic systems are not an option when operating in the vasculature. Among the medical imaging techniques widely used in clinics and hospitals and capable of imaging inside the human body, X-ray or Computed Tomography (CT) scan, and MRI are presently thought to be the two main imaging modalities that offer potentially acceptable spatial resolution and speed for such applications. Other imaging modalities such as ultrasound and Positron Emission Tomography (PET) could be used as complementary means with X-ray and/or MRI.

In the planning phase where the trajectory (navigation path) of the nanorobots from the release site to the target site must first be determined, it is first necessary to be able to image the planned route, *i.e.* the blood vessels that will be traveled. In the medical field, angiography is well known as the main technique to perform such a task. Although presently X-ray Digital Subtraction Angiography (DSA) has the best spatial and temporal resolution for imaging blood vessels, the method offers poor soft-tissue information and exposes the patient and the medical staff to high radiation. These two disadvantages do not apply with MRI. Because of that, substantial research efforts to improve Magnetic Resonance Angiography (MRA) beyond DSA are underway. Despite substantial recent progress in MRA, DSA is still the preferred and most appropriate technique to image the vascular network.

A simple example of the use of X-ray for the navigation path planning is depicted in Fig. **5**. In this particular case, the carotid artery of a swine is imaged using contrast agents (top right of Fig. **5**). Then the acquired image is filtered to remove all image artifacts. The filtered version provides an image of the blood vessel with sharp edge and no artifacts. This filtering facilitates the path planning while avoiding any possible confusion by providing a clear picture of the vessel to be navigated. Then waypoints are plotted along the vessel to be navigated. These waypoints represented by dots superposed on the filtered image of the blood vessel (here the carotid artery) is used to indicate to the computer (controller) the trajectory to be followed by the microscale nanorobots. In this particular *in vivo* experiment [14] which was a first demonstration proving the possibility of navigating an untethered object in the blood vessel of a living animal, a 1.5 mm in diameter ferromagnetic bead was servo-controlled by an external computer at a rate exceeding 20Hz (*i.e.* more than twenty tracked position + control + generation of corrective propulsion gradient cycles per second), navigating back and forth along the waypoints located in the carotid artery at an average velocity of 10cm/s. As depicted in Fig. **5**, the tracked position of the bead acquired using MRI, is superposed on the X-ray image taken previously and aligned using registration techniques. As shown, the MR-tracked displacement of the bead follows the carotid artery visible next to the spinal cord to the right and to one of the legs on the left of the image.

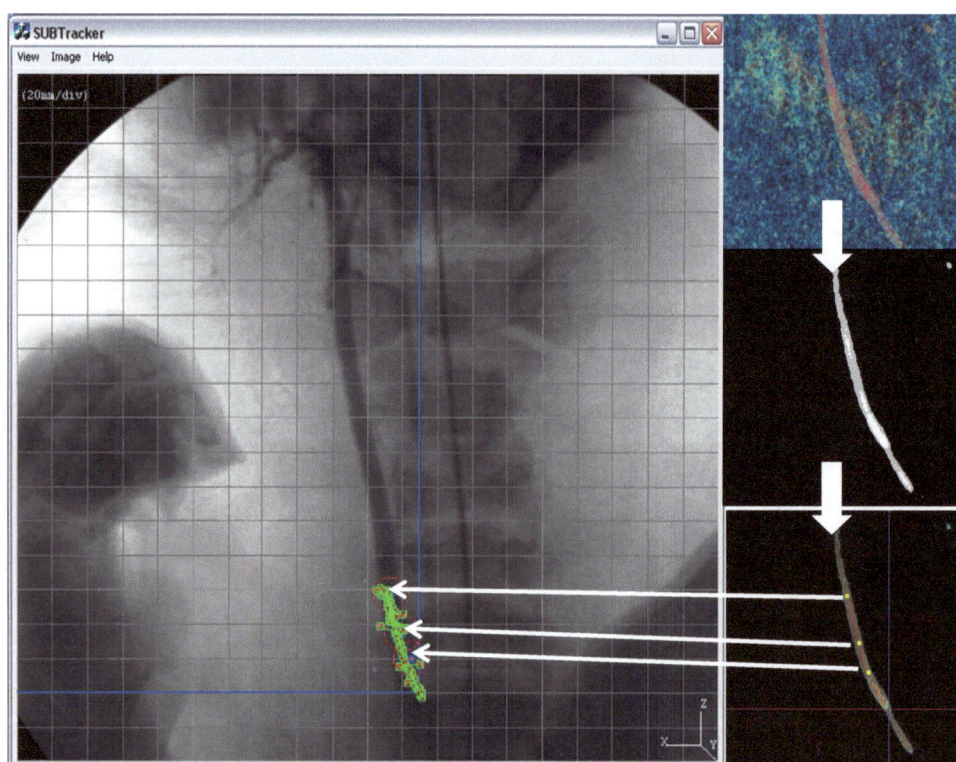

**Figure 5:** (Top-right) Image of the carotid artery of a living swine before filtering; (Middle-right) Filtered image of the artery defining the accessible route; (Bottom-right) Waypoints superimposed on the artery to indicate the trajectory to be followed; (Left) MR-tracked position of a 1.5mm magnetic bead being automatically navigated along the waypoints in the carotid artery of a living swine at an average velocity of 10cm/s. (Adapted from [14])

But DSA may be effective during the pre-navigation/planning phase of the microscale nanorobots but not necessary for gathering real-time tracking positions during travel. The spatial resolution of X-ray is typically better that clinical MRI but is still not sufficient to detect a single or a few nanorobots in the microvasculature. To be detectable, a single aggregation of such nanorobots must be sufficiently large, *i.e.* at or larger than the spatial resolution of the system. Such aggregation would then be detectable similarly to contrast agents. But although DSA provides a good image of larger vessels, the technique cannot image smaller diameter vessels including the capillaries. Similarly, X-ray will easily detect and track larger robots and objects (such as the 1.5 mm in diameter bead mentioned previously) but will fail to detect and track microscale nanorobots such as TMMC in smaller diameter vessels (*e.g.* small diameter arterioles and capillaries). This is because the volume limitation when in the microvasculature will prevent the implementation of sufficiently large aggregates to be detectable using X-ray. One solution to this limitation in medical imaging is to amplify a signal (artifact) from the nanorobots themselves to a size sufficiently large to be detectable by the medical imaging system. This is shown in Fig. **6**. When placed in a high intensity homogeneous magnetic field such as the one found in the bore of clinical MRI scanners, a ferromagnetic object will create a net field inhomogeneity. This is depicted on top of Fig. **6** where such artifact is much larger (at least 50 times larger) than the object itself. That means that an object such as a microscale robot made or containing a sufficient amount of iron for instance (although other materials could also be considered), could have overall diameter allowing it to travel in part the microvasculature while being detectable by MRI. For instance, a voxel with edges of 0.5mm could theoretically detect and track an object such as a microscale robot with a diameter of less than 10μm.

Indeed, the local magnetic field distortion from a magnetic spherical particle for instance can be found at a point *P* of coordinate *r(x, y, z)* by that of a magnetic dipole as

$$\vec{B}'(P) = \frac{\mu_0}{4\pi}\left(3\frac{(\vec{m}.\vec{r})\vec{r}}{r^5} - \frac{\vec{m}}{r^3}\right),$$

(1)

where $\mu_0 = 4\pi \times 10^{-7}$ H·m$^{-1}$ is the permeability of free space such that for a uniformly magnetized object, the dipolar magnetic moment (A·m$^2$) depends on the magnetization saturation of the material and is given by

$$\vec{m} = \frac{4}{3}\pi a^3 \vec{M}_{SAT},$$

(2)

where *a* here represents the radius (m). Accordingly, the above equations show that a magnetic robot undetectable under X-ray could potentially become visible and hence trackable using MRI if the amplitude of distortion of the magnetic field away from the microscale robot and at the edge of a voxel (3D pixel that defines the spatial resolution of the MRI system) would be equal or above the sensitivity level of the MRI scanner.

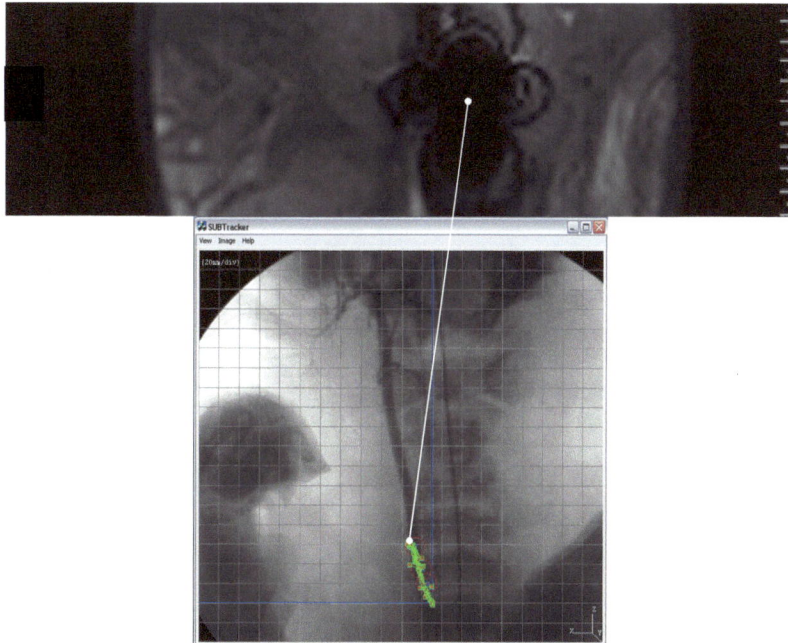

**Figure 6:** (Top) Image artifacts created by the ferromagnetic beads due to inhomogeneity of the magnetic field in a clinical MRI scanner; (Bottom) The MR-tracked position computed using the MS-SET method.

Although the image artifact created by an entity with dimensions small enough to be undetectable with X-ray can be larger than a voxel, then allowing its detection with MRI, computing the 3D position of the object or microrobot within such a relatively large artifact with sufficient accuracy and in real-time as depicted in the lower image of Fig. **6**, requires a special technique to be developed for such application. Fortunately, such a technique has been developed for this purpose and it is referred to as Magnetic Signature Selective Excitation (MS-SET) [17].

The method MS-SET is based on the application of a Radio-Frequency (RF) excitation tuned to the frequency of the desired magnetic equipotential curves. Encoding the positions of the excited spins is then performed as in traditional imaging, *i.e.* through the application of a readout gradient. A projection image of the magnetic equipotential is then obtained whose shape depends on the excitation frequency as well as in the dipole magnetic signature. The projection along the read axis corresponds to the central line of the k-space along the same axis. Tracking in a 3D volume requires only three k-space lines (one for each orthogonal axe) resulting into sufficient real-time performance for such an application. The measured displacement corresponds to the strongest correlation between each subsequent position and the first one. In previous experiments, an average precision in the order of 160μm (approximately half a voxel precision) have been obtained using a clinical MRI scanner. In the dynamic regime, the precision in tracking achieved was in the order of 600μm. These preliminary results suggest that tracking is indeed possible providing a good estimate of the position of the nanorobots but with limitation in determining the exact vessel being traveled when in the arterioles or the microvasculature considering the distance separating adjacent vessels in these regions.

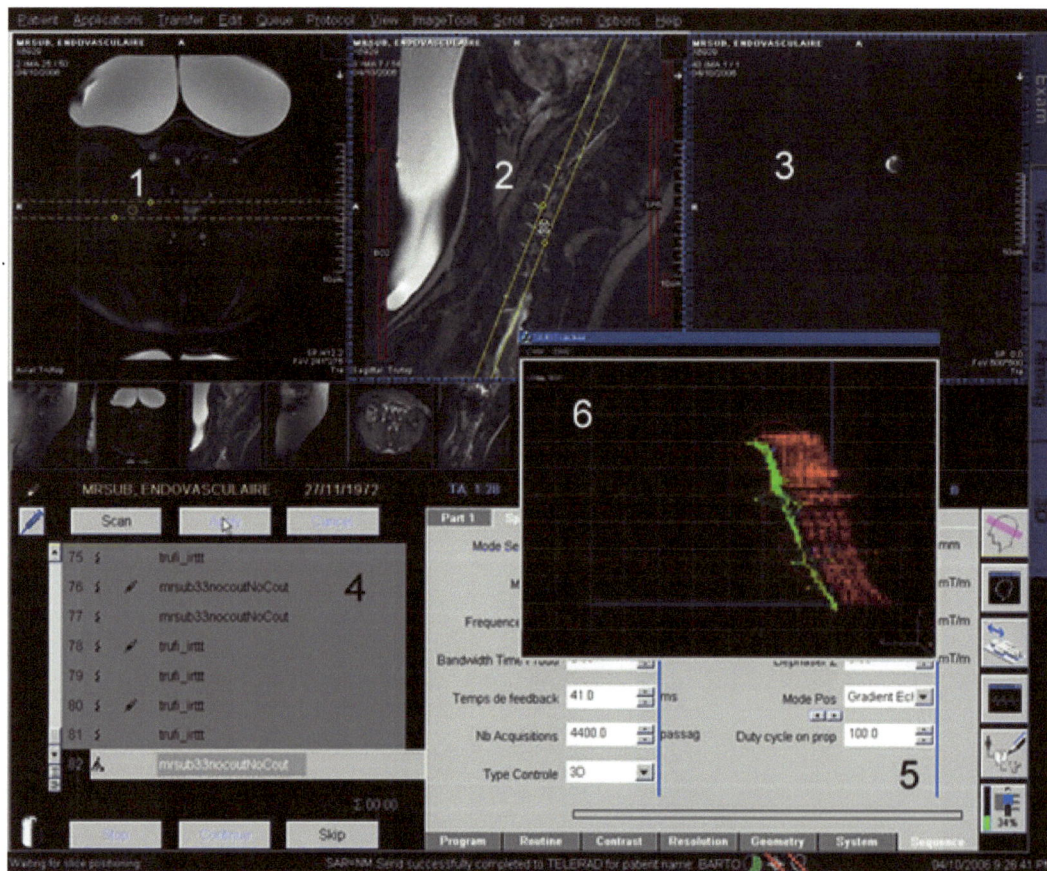

**Figure 7:** (1 and 2) MR-images taken to make some measurements prior to the real-time navigation phase; (3) MR-image taken prior to the real-time navigation phase to determine if there is any objects creating a signal that could be confused with the signal created by the object to be navigated; (4 and 5) Windows for entering various parameters; and (6) trajectory of the navigated object being displayed in real-time. (From [9])

But there is another important issue. Although the net field inhomogeneity created by the microscale nanorobots allows them to be detected and tracked in the body using MRI, such inhomogeneity also prevents the imaging of the surrounding tissue within such artifact. Considering that MR-imaging relying on several slices takes a considerable time relative to the real-time closed-loop control frequency being used, complete imaging in the working volume is unlikely to be taken during the navigational phase but in the non real-time phases such as during the pre-operative phase dedicated to planning. Windows 1 and 2 in Fig. **7** are examples of that. Therefore, although complete imaging is unlikely during the real-time navigational phase of the nanorobots, gathering some partial image information may still be suitable for registration purpose. In these cases, the detectable overall size of the artifact produced by the ferromagnetic material embedded in the microscale nanorobots must be minimized.

But the fact that same ferromagnetic material is also used for propulsion/steering put some constraints. More specifically, if the volume of ferromagnetic material is reduced to minimize the size of the artifact used for tracking, then the propulsion force induced will be reduced as well. When operating in larger diameter vessel, the accuracy in registration becomes less important and as such, a larger artifact from a larger volume of ferromagnetic material to achieve higher propulsion force can be tolerated. Using MS-SET, once the position is determined, an icon representing the position of the navigable entity is plotted by the computer on top of the planned trajectory as depicted in Fig. **8**.

In smaller diameter vessels, an aggregate of small entities or microscale nanorobots is used with each one having the right amount of magnetic material to produce an image artifact of adequate size while providing sufficient propulsion/steering force. In this case, dipole-dipole interactions between each microscale nanorobots can be exploited to increase the propulsion/steering force well beyond the level obtained with a single nanorobot. This is explained in more details in the next section.

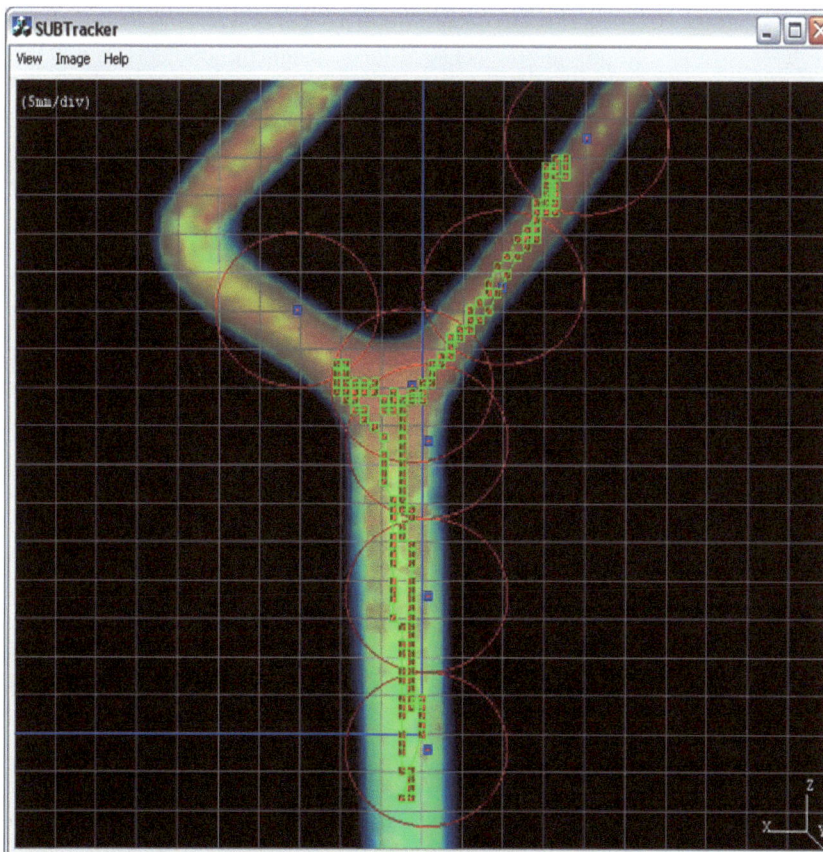

**Figure 8:** Position of an untethered ferromagnetic bead being displayed in real-time during back-and-forth navigation along waypoints at vessel bifurcation in a phantom mimicking the human vasculature.

## 6. INDUCED PROPULSION AND STEERING

For operations performed by untethered devices in the human body, propulsion/steering techniques based on magnetism are presently the most promising. The simplest approach is with the use of a permanent or electro-magnet such as the one depicted on the left in Fig. **3**. The latter creates a non-uniform magnetic field of increasing magnitude towards the magnet used to produce the magnetic field for propulsion. An orthogonal configuration consisting of three electromagnets could theoretically control the displacement of the untethered robot in a 3D space within the human body. But controlling the displacement of such robot would be an extremely difficult task that would require extremely fast feedback corrective actuations, most probably beyond what could be possible with present technologies.

To simplify the issue of controlling the robot within the required spatial and temporal constraints, magnetic gradients that are constant throughout the working space can be used [18]. In this case, the magnetic force $F_M$ (N) (Eq. 3) induced in the magnetic core with a volume $V$ (m$^3$) embedded in the robot and produced by magnetic gradients (T m$^{-1}$), depends on the duty cycle $R$, *i.e.* the percentage of time per navigation closed-loop control cycle being dedicated to propulsion, and the volume magnetization of the core material $M$ (A m$^{-1}$).

$$\vec{F}_M = R \cdot V \left( \vec{M} \cdot \nabla \right) \vec{B} \quad .$$

(3)

To achieve the highest propulsion force density, *i.e.* the greatest propulsion force for a given magnetic core's size, the magnetic material embedded in the robot must reach a saturation level ($M_{SAT}$) referred to as magnetization saturation. Interesting enough is the fact that a homogeneous field of 1.5T or higher as the one found in the bore of a clinical MRI scanner is sufficient to achieve such magnetization saturation for the two main material types of interest for this application namely, iron-oxide and iron-cobalt. More interesting is the fact that the orthogonal gradient coils implemented in a clinical MRI scanner and used for image slice selections, can indeed provide the constant linear magnetic gradients in a 3D working space that is large enough to accommodate a human adult. Because gradients used for propulsion/steering would interfere with the image slice gradients, both types of gradients from the same coils would typically be generated in a time multiplexed fashion, hence leading to a reduction of $R$ (where $R = 1$ indicates that the coils are continuously used for propulsion purpose without imaging/tracking) when imaging/tracking must be performed. But a relatively large magnetic core will create an image artifact in proximity of the robot that will prevent MR imaging of surrounding tissues. Accordingly, as described in Eq. 1 and Eq. 2, if maximum magnetization is required for maximum induced magnetic force for the robot, then as suggested in Eq. 2, decreasing the size of the image artifact can be achieved by decreasing the volume of the magnetic core accordingly. But by doing so, the resulting magnetic force (Eq. 3) will decrease proportionally. One solution to this issue is a variation of Eq. 3 and which is described in Eq. 4.

$$\vec{F}_{MR} = R \cdot \left( \sum_n V_n \right) \left( \vec{M} \cdot \nabla \right) \vec{B}, \quad \sum_n V_n \leq V_R \quad .$$

(4)

Unlike Eq. 3 which applies to a single large magnetic entity for each robot, Eq. 4 defines the magnetic force induced on a robot which has $n$ magnetic cores (*e.g.* MNP) which can be of identical or dissimilar sizes where the total volume of magnetic material embedded is smaller or equal (typically smaller) that the total volume of the robot itself. Hence, to cope with the velocity of the blood flow, such synthetic entity or microscale synthetic robot would then have a magnetophoretic velocity expressed as

$$\overrightarrow{v_M} = \frac{\vec{F}_{MR}}{f} \quad .$$

(5)

The variable $f$ is the friction factor which depends on the geometry of the robot being navigated. When such ferromagnetic-based robot is inserted in a high magnetic field such as the DC homogeneous field inside the bore of a MRI scanner and known as the $B_0$ field (typical values for clinical scanners are 1.5T or 3.0T), such robot will align to the lines of this magnetic field and maintain the same orientation while operating in the MRI bore. Hence, a spherical external shape of such robot would be a good choice, resulting in a constant friction factor that would be independent of the direction of any navigated blood vessels relative to the $B_0$ field. In such a case, the friction factor for such untethered microscale spherical synthetic robot would be computed as

$$f = 3\pi \eta d_R .$$  (6)

In Eq. 6, $\eta$ is the viscosity of the medium (*e.g.* blood), and $d_R$ is the diameter (m) of the spherical robot.

But imaging gradient coils are designed for fast slew rates which limits in a clinical setting, the maximum amplitude that can be generated and the time at which, a higher amplitude can be sustained. Indeed, typical clinical MRI scanners provide maximum gradients of approximately 40mT/m. This value is sufficient for navigating a larger robot (with enough magnetic material) in larger diameter blood vessels such as the arteries, but becomes insufficient for a smaller robot that must travel in smaller diameter vessels. As such and as depicted in Eq. 4, it appears that the most obvious solution is to increase the effective volume of magnetic material embedded accordingly to compensate for such limitation. But again, such increase in volume is limited by the volume of the robot itself with the latter being limited by physiological constraints and most particularly by the smallest diameter blood vessel being transited before reaching the targeted site.

One solution to this limitation is to increase the effective volume of magnetic material while being able to navigate in various vessels geometries and diameters by replacing a single larger robot by an aggregation of smaller robots. Indeed, sufficiently high coupling forces lead to more effective propulsion/steering forces. When placed in a high intensity magnetic field such as the $B_0$ field in a MRI scanner, each ferromagnetic-based microscale robot will generate a local magnetic dipole. When a large number of such robots are injected as a whole, the dipole-dipole interaction energy $E_D$ between neighboured robots while maintain them in an aggregated state. This is depicted in Eq. 7 for two robots.

$$E_D = -\frac{\mu_0}{4\pi}\left( \frac{(\mu_A \cdot r_{AB})(\mu_B \cdot r_{AB})}{r_{AB}^5} - \frac{\mu_A \cdot \mu_B}{r_{AB}^3} \right), r_{AB} = \|r_{AB}\|$$  (7)

Equation 7 can be used to estimate the dipole-dipole interaction energy $E_D$ between two neighboured robots A and B with respective dipoles $\mu_A$ and $\mu_B$. Such interaction depends not only on the relative orientations of the dipoles but also on their orientation with respect to the vector $r_{AB}$ joining the center of the two dipoles.

With strong interactions, such aggregate will effectively increase the volume of magnetic material and allow higher magnetophoretic velocities while a lower level of dipole-dipole interactions will allow such aggregate to reconfigure when transiting between various vessels geometries such as transiting from larger to smaller diameter vessels. On the other hand, too strong dipole-dipole interactions may lead to an unexpected embolization (*e.g.* when transiting from larger to small diameter vessels or at vessel bifurcations) prior to reach the planned targeted area whereas too weak interactions may lead to unexpected breakages resulting in a loss of targeting efficacy from independent smaller aggregates. Therefore, the right compromise in the level of dipole-dipole interacting forces must be set during the synthesis of such microscale robots.

When such dipole-dipole interactions occur in a MRI scanner, the lowest energy configuration will govern the geometry of such aggregates or clusters. Such lowest energy configuration would then correspond to the two magnetic moments aligned head-to-tail. In other words, neighboured (*i.e.* close enough for $E_D$ to be non-negligible) microscale robots will tend to form columns of aggregates with the elongated axe being oriented in the direction of the $B_0$ field. This phenomenon is depicted in Fig. 9.

The *in vitro* experimental results [19] depicted in Fig. 9 in a phantom mimicking realistic human physiological data show the steering efficacy of such aggregates for magnetic gradient up to 400mT/m being approximately the maximum gradient that can be implemented as for am insert installed in the bore of a clinical MRI scanner for interventions in a human adult.

These tests and further tests performed *in vitro* and *in vivo* also showed the importance of using not only gradients up to 400mT/m but also the potential advantages of embedding superparamagnetic instead of permanent (hard) or soft ferromagnetic nanoparticles in each nanorobot. Superparamagnetism occurs in magnetic materials composed of very small crystallites and occurs for particles with overall dimensions in the nanometer-scale. These MNP will be superparamagnetic below a threshold dimension that depends on the type of materials. For example, a Fe-based MNP becomes superparamagnetic at sizes below 25 nm.

Superparamagnetic MNP are extremely important components to be embedded in these nanorobots because they can be magnetized with an external magnetic field without retaining any residual magnetism in the absence of the applied field, preventing the risk of uncontrolled aggregations during the medical intervention due to dipole-dipole interactions. Although dipole-dipole interactions to create aggregates are important during the navigation phase from the catheterization boundary or release site to the target or planned embolization site, the formation of an aggregation is not suitable during the injection since it would prevent the release of the nanorobots in the blood to become jammed in the release catheter (or a syringe) when the patient placed in the MRI bore for immediate computer-based navigation upon release. After the target is reached and the patient is removed from the high magnetic field of the MRI scanner, non-superparamagnetic such as hard and soft ferromagnetic particles would cause unwanted embolization and further potential complications. As such and because of the importance of a high magnetic field when using superparamagnetic material, makes the MRI scanner with its $B_0$ field, an ideal platform in this respect when it is compared to other medical imaging platforms that do not provide such high field and which include well known systems such as X-ray, CT, and PET.

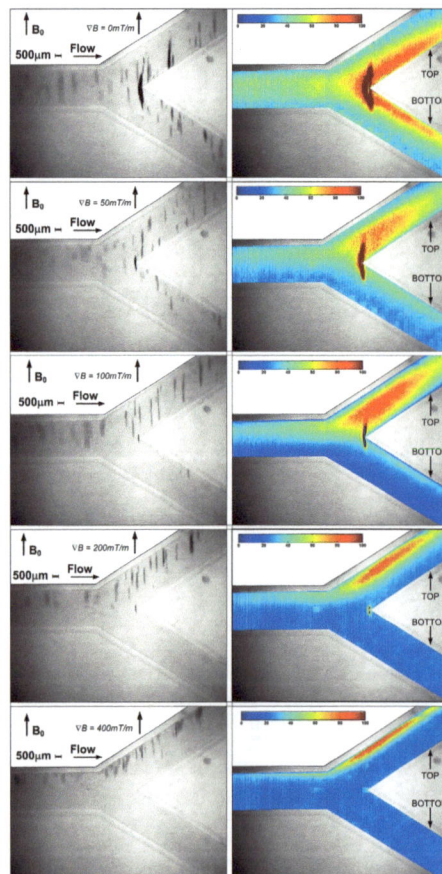

**Figure 9:** Effect of magnetic gradient amplitudes on the steering ability at vessel bifurcation.

## 7. SELF-PROPULSION AND INDUCED STEERING

A self-propelled robot is defined here as a robot that can propel itself without any external source such as the gradient fields mentioned in the preceding section. Gradient-based propulsion is limited mainly by the magnetization saturation level of the material, the maximum gradient that can be provided at the human scale, and the volume of magnetic material as depicted in Eq. 4. Although the effective volume of magnetic material can be somewhat increased with an aggregation caused by dipole-dipole energy (Eq. 7), space constraints imposed by smaller diameter vessels and shorter distance between vessel bifurcations especially in the microvasculature prevent such effective volume to be large enough to allow sufficient propulsion/steering force to be induced for targeting purpose. As such, self-propulsion could be an efficient alternative means of propulsion in these regions. Ideally, such self-propelled systems should have dimensions appropriate to propel themselves in the smallest diameter vessels including the angiogenesis network that must be travelled in order to reach a tumoral lesion and as such, wall retardation effect [20] must be taken into account. The wall retardation effect is expressed as

$$\frac{v}{v_\infty} = \left(\frac{1-\lambda}{1-0.475\lambda}\right)^4 ,$$

(8)

where $v$ and $v_\infty$ are the velocity of the spherical robot in the blood vessel (here the capillary) and in open space respectively. The variable $\lambda = d/d_C$ is the ratio between the diameter of the robot or the cell of and the diameter of the blood vessel or capillary being traveled respectively. For a spherical synthetic nanorobot propelled by magnetic gradient as described in the preceding section, maximizing $d$ is a must to increase the volume of magnetic material and hence the magnetophoretic velocity required to cope with the blood flow velocity in the respective regions being traveled. On the other hand, increasing the overall size well beyond $\lambda = 0.5$ would cause the magnetophoretic velocity to be decreased by wall retardation effect. In order words, the gained velocity achievable with a larger diameter spherical magnetic nanorobot would not be possible for nanorobots with diameter exceeding approximately half the diameter of the blood vessel being traveled. For the smallest capillaries found in a human, this optimal diameter would correspond to approximately 2μm. From the previous equations and especially Eq. 4 and Eq. 5, one can deduct that gradient-based propulsion for such small diameter spherical synthetic robot is well beyond technological limits especially when operating in regions of the body such as the torso (mainly due to the required inner diameter of the coils to accommodate a human adult).

Instead of considering a synthetic or an artificial implementation, one other approach would be to consider what nature provides. For instance, a flagellated bacterium could be used as natural self-propulsion system for a miniature robot. In fact, being able to control the motion of such a cell could define it as a natural self-propelled robot. As such, Magnetotactic Bacteria (MTB) has been proposed and considered as means of actuation, propulsion and steering for the implementation of microscale robots (4) in applications performed in the human microvasculature. Among all species of MTB, the MC-1 cell depicted in Fig. **10** as been selected as the best candidate for such medical applications.

As depicted on the left side of Fig. **10**, the MC-1 cell is round in shape with a diameter of ~2μm, which is adequate according to Eq. 8 for travelling in the smallest diameter vessels of the human microvasculature including the angiogenesis network. But for the design of an efficient self-propulsion system to be used in the microvasculature, the Reynolds number $Re$ must be considered. In fluid dynamics, $Re$ is a dimensionless number that gives a measure of the ratio of inertial forces to viscous forces. Besides offering insights about how the self-propulsion system should be designed, it is also used to characterize different flow regimes (such as laminar versus turbulent flow) and hence the type of fluid environment that the robots will operate when in the vascular network. Laminar flow for instance will occur at low Reynolds numbers with viscous forces being dominant. The latter will be characterized by smooth and constant fluid motion. Turbulent flow regimes on the other hand will occur at high Reynolds numbers. In the latter environment, inertial forces will dominate and as such, will tend to produce random eddies, vortices and other flow instabilities. For the blood flow in a capillary of diameter $d_C$, $Re$ can be computed as

$$\mathrm{Re} = \frac{\rho V d_C}{\mu}.$$

(9)

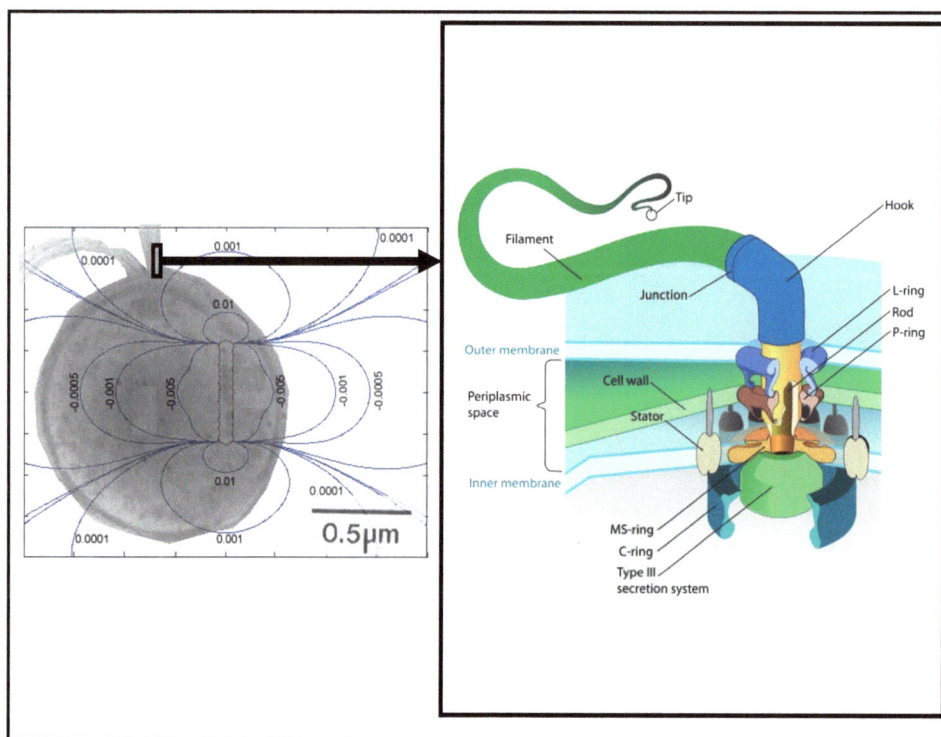

**Figure 10:** (Left) Image of the MC-1 cell; (Right) Schematic of the bacterial molecular motor with the flagellum attached (image from [21]

In the previous equation, $\rho$ is the density of blood (kg/m$^3$), $V$ is the mean blood velocity (m/s) at the particular capillary being traveled, and $\mu$ is the dynamic viscosity of the blood in the capillary (Pa · s or N · s/m$^2$ or kg/m · s). From Eq. 9, it becomes obvious that a microscale robot travelling in the human microvasculature and in particular in the capillaries would operate at very low Reynolds hydrodynamic conditions being dominated by viscous forces. It is well know that in a viscous medium, a propulsion system relying on a flagellum such as the one depicted in the right section of Fig. **10** would offer much higher performance than the typical propeller used in our typical macro-scale environment characterized by higher Reynolds numbers. As depicted in the figure, this flagellum is connected to a molecular or nanoscale motor embedded in the flagellated bacterium. This motor measures less than 300 hydrogen atoms across, and has its macroscale counterpart, it has a stator and a rotor attached to the flagellum that allows propulsion with 360-degree revolutions. This rotary engine composed of proteins is powered by a flow of protons which makes it particularly attractive since as mentioned previously, electrical power is an important factor that limits other robots to be miniaturized at such a scale. The MC-1 MTB for instance is not only non-pathogen but with its two flagella bundles, the cell can swim at an average of 100 body lengths and for a percentage of them, peak velocities of approximately 300μm/s have been recorded (approximately ten times the swimming velocities of many other species of flagellated bacteria). This translates to a propulsion trust force between approximately 4.0 and 4.7pN compared to 0.3-0.5pN for many other species of flagellated bacteria including well known species. But what makes the MC-1 bacterium attractive as a natural microscale nanorobot is the fact that unlike most flagellated bacteria that are based on chemotaxis to detect nutrient gradients and hence influence their motility [22-24], the direction of displacement this specie of bacteria can be controlled remotely in the human body by an external computer. This is possible because MTB and in particular the MC-1 bacterium unlike other species of flagellated bacteria, contains as depicted in Fig. **11** (lower left corner), a chain of magnetosomes which are membrane-based magnetite (Fe$_2$O$_3$) nanoparticles that act like a nanoscale magnetic compass.

Indeed, although influenced by other means such as chemotaxis and aerotaxis, MTB are also influenced by magnetotaxis [25-27]. Previous experiments showed that when the MC-1 cells for instance are subjected to a magnetic field's intensity slightly higher than the earth's magnetic field of 0.5 Gauss, the directional motion of these MTB can be influenced in a precise manner by exploiting magnetotaxis without noticeable motion effects from other means. Since a directional electromagnetic field can be produced by an electrical current, an interface between modern computerized platforms and MTB can be established, allowing full steering controllability from algorithms coded as computer software [28]. Directional control of the MC-1 cells is done by inducing a directional torque from a 3D directional magnetic field. Because propulsion odes not relies on an external source since the bacteria are self-propelled, the magnetic field required is relatively very low in comparison since induced steering require a very low field intensity that can be technologically achieved at the human scale.

The directional control is generally generated from computer algorithms. These algorithms take into account the swimming motion behavior of the MC-1 cells when they encounter various obstacles such as the ones expected in the microvasculature including the angiogenesis network, and modulates a magnetic field outside the MRI scanner and from another special platform in a way to maximize the chance of the bacteria to reach the tumor. This is critical since the closed-loop control method depicted in Fig. **4** cannot be applied in this case. The main reason is that there is no medical imaging modality with sufficient spatial resolution capable of imaging the tiny blood vessels in the microvasculature. Hence, the fact that the MTB cannot be controlled in the bore of a clinical MRI scanner is not an issue. Therefore, since no specific trajectory can be plotted over the blood vessels, the controller has no reference to compute and to apply corrective actions. In fact, only the tumor and the position of the aggregate of the MC-1 cells can be imaged when inside the vasculature. Indeed, interesting enough is the fact that each single magnetic domain $Fe_2O_3$ nanoparticles (each with a diameter of approximately 70nm) in the MC-1 cell not only provide directional control but also act like a MRI contrast agent by creating a small field inhomogeneity (see Eq. 1 and Eq. 2) that can be detected (with a sufficient density of MTB) with a clinical MRI scanner. This is depicted in the lower right portion of Fig. **11**.

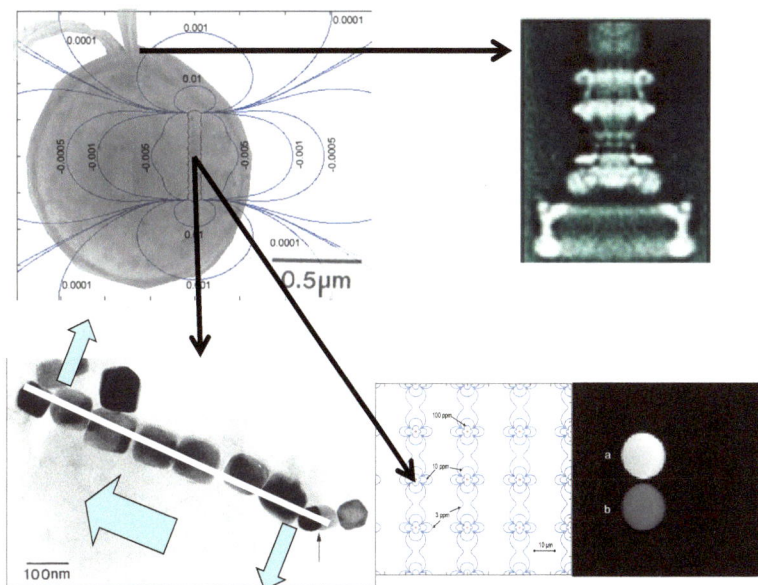

**Figure 11:** (Top-left) The MC-1 flagellated bacterium; (Top-right) Photograph of one of the molecular motor; (Bottom-left) Chain of magnetite nanoparticles called magnetosomes being used for directional control by computer; (Lower-right) Magnetosomes distorting the $B_0$ field for MRI-tracking.

As depicted in the lower left portion of Fig. **11**, a clinical MRI scanner can, with an appropriate imaging sequence, evaluate the concentration (density) of MTB within a region of interest (ROI) which would typically be the targeted region. The concentration can be evaluated from the change of intensity of the image caused by the magnetosomes. This is important since it allows the medical staff to monitor and evaluate the concentration of therapeutic agents delivered within some level of accuracy provided that the MC-1 cells were carrying therapeutic agents. This in turn can provide data about the percentage of agents that did not reached the target and hence the level of secondary toxicity expected. In turn, this level of toxicity may help the medical staff to decide if further attempts to target the tumor can be done without major risks for the patient.

## 8. POLYMORPHIC MICROSCALE NANOROBOTS

A simplified diagram of the travel path of the nanorobots is depicted in Fig. **12**. Although a single relatively large magnetic microscale nanorobot such as a TMMC could travel in larger blood vessels, an aggregate of smaller diameter TMMC being propelled and steered by magnetic gradients generated by an upgraded MRI scanner could travel from a catheter release site to deliver therapeutic agents at an embolization site indicated in Fig. **12** by arrow 1. Beyond such boundary, gradient-based propulsion would be limited by the small effective volume of magnetic material. On the other hand, self-propelled MC-1 bacteria are very effective for travelling in the region indicated by arrow 3 with more difficulty in the region indicated by arrow 2. This is due to the fact that the average blood flow is typically higher in larger diameter vessels. As such, Therapeutic MTB (TMTB), *i.e.* MTB carrying therapeutic loads (*e.g.* by attaching nanoparticles containing therapeutic agents to the cell using antibodies) would be embedded in special gradient-propelled carriers and released at an embolization site. Such antibodies specific for this application, have already been developed and the synthesis of carriers capable of transporting MTB are presently under investigation. Although using TMMC alone could provide an effective tool to fight many solid tumors through chemo-embolization, the use of other approaches to get closer to the solid tumor could improve therapeutic efficacy further in two ways.

The first one would be to perform chemo-embolization closer to the solid tumor while the second would be to reduce the blood flow in larger capillaries in order to make the swimming velocity of the MC-1 cells more effective towards targeting the solid tumor; and one powerful concept to achieve these two approaches will be with the use of polymorphic microscale nanorobots.

A polymorphic microscale nanorobot is one that can change its shape. In its simplest form, such polymorphic microscale nanorobot would change volume within the micrometer-scale when triggered, typically from an external computer. For instance, once at the embolization site indicated by arrow 1 in Fig. **12**, such robot could reduce its diameter to allow it to drift with the blood flow deeper in the microvasculature and then to stop embolization by expanding its volume (diameter).

To get to the same embolization site, non-polymorphic TMMC would have to be initially implemented with a smaller diameter and hence with a smaller volume. With a smaller volume, less magnetic material would be embedded, leading to a lower magnetophoretic velocity. Such velocity or force induced would become insufficient to stop the TMMC against the blood flow and would most likely continue its course and reach the systemic blood circulatory network, hence increasing at the same time the level of secondary toxicity while decreasing the therapeutic index. On the other hand, a polymorphic TMMC with the same diameter (*e.g.* ~50μm) and charged with the same quantity of the same magnetic material, would also reach the same initial embolization site (*e.g.* arrow 1 in Fig. **12**) but would continue its course using the blood flow as propulsion (but without steering) with a decrease of its volume and then stop deeper in the vasculature at a second embolization site (*e.g.* arrow 2 in Fig. **12**) due to an expansion of its volume.

A first implementation (29-30) relies on MNP embedded in N-isopropylacrylamide (NIPA) hydrogel. As for non-polymorphic TMMC, the embedded MNP in polymorphic entities are used for propulsion, MR-tracking, and are responsible for the formation of aggregates through dipole-dipole interactions. But in polymorphic implementations, the embedded MNP are also used to trigger the change in volume from commands sent by a computing platform. NIPA reduces its volume when heated and expanded back to its original volume when it cools down. This discontinuous and reversible mechanism caused by internal structures changes is due to the presence of hydrophobic groups occurring at the Lower Critical Solution

Temperature (LCST). This LCST can be set slightly above body temperature by copolymerization with hydrophobic monomer like acrylic acid [23-24]. Adjusting the LCST just a few degrees (*e.g.* 3-4 degrees) above the internal body temperature of 37°C, could also allow targeted hyperthermia leading to enhanced therapeutic efficacy with a local increase of internal temperature without damaging the viable cells. At temperatures below the LCST, hydrogen bonds between the hydrophobic groups of polymer chain are dominant, *i.e.* that the dissolution of water has increased. On the other hand, at the LCDT hydrogen bonds become weaker and hydrophobic interaction become stronger, resulting in a change of the internal structure with water being expelled out of the hydrogel.

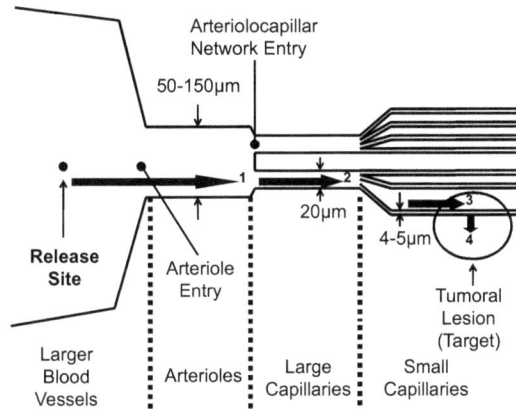

**Figure 12:** Simplified diagram showing the diameter of the blood vessels between the catheterization boundary and the tumoral lesion.

The increase from the internal body temperature to the LCDT is done from heat generated by the MNP when placed in an AC magnetic field, a principle known in hyperthermia. It is well known in the field of hyperthermia and considering the viscosity of the hydrogel that a single domain MNP would create more heat with a Specific Absorption Rate (SAR) obtained from Néel and Brownian relaxation. This can be computed as

$$SAR = \frac{V(M_S H \omega \tau)^2}{2\tau kT(1+\omega^2\tau^2)},$$

(10)

where

$$\frac{1}{\tau} = \frac{1}{\tau_N} + \frac{1}{\tau_B} \cong \frac{1}{\tau_N} = \frac{2\sqrt{\frac{KV}{kT}}}{\sqrt{\pi}\,\tau_0\,e^{\left(\frac{KV}{kT}\right)}}.$$

(11)

In Eq. 10, $V$ is the mean volume of the MNP, $H$ is the magnitude of the magnetic field, $\omega$ is the angular frequency, $KV$ is the anisotropy energy (K = $1.35 \times 10^4$ J/m$^3$ for Fe$_3$O$_4$), and $kT$ is the thermal energy. As computed by the above equations, maximum loss and hence heat is achieved when $\omega\tau \approx 1$. With a magnetic field at frequency $\omega/2\pi = 145$ kHz for instance, optimum relaxation time $\tau = 1.0976 \times 10^{-6}$ s, which corresponds to a MNP with a diameter of ~17nm. Other diameters are possible by adjusting other parameters (*e.g.* the frequency) in the interventional platform.

## 9. DIRECT-PASSIVE (DP) AND DIRECT-ACTIVE (DA) TARGETING

With the help of polymorphic microscale nanorobots, the flagellated bacterial nanorobots could become more effective at targeting the tumoral lesion indicated by arrow 3 in Fig. **12**. But to get deeper in the tumor, the segment indicated by the arrow 4 in Fig. **12** would have to be taken. Normally this would be done only through the capillary wall. The capillary wall is a one-layer endothelium that allows gas and small molecules to pass through but not bacteria including the MC-1 cell with an overall diameter of ~2µm. Unfortunately, unlike normal blood vessels, tumor vessels are unusually leaky and such leakiness may contribute to allow the MC-1 cells to reach deeper regions in the tumor. For instance, some investigations done in tumors [31] have revealed that some branched cells can be separated by intracellular openings in the range of 0.3-4.7µm. But at present, more investigations are needed to determine how far such MC-1 cells under the guidance of a computer can go. Nonetheless, if the MC-1 cells could deliver nanoparticles capable of passive or active targeting, the therapeutic efficacy could potentially be increased further.

Presently, tumor targeting with MNP uses passive or active strategies. Passive targeting occurs as a result of extravasation of the MNP at the tumor where the microvasculature is hyper-permeable and leaky. On the other hand, active targeting is based on the over or exclusive expression of different epitopes or receptors in tumoral cells, and on specific physical characteristics (*e.g.* vectors sensitive to physical stimuli such as temperature, pH, electric charge, light, sound, magnetism, *etc.* and conjugated to drugs). Furthermore, active targeting may also use over-expressed species including low molecular weight ligands (such as folic acid, thiamine, or sugar), peptides, proteins (such as transferring, antibodies, or lectins), polysaccharides (such as hyaluronic acid), polyunsaturated fatty acids, peptides, DNA, to name but the main ones.

Passive and active targeting can be relatively efficient for the segment indicated by arrow 4 in Fig. **12**. But the main issue is to get close enough and as such, being able reach the tumoral regions directly with the nanorobotic methods described earlier while avoiding the systemic circulation may enhance further the therapeutic efficacy. In turn, nanorobotics may benefice from existing passive or active targeting approaches, especially for the last travelling segment, leading to new paradigms that could be referred to as Direct-Passive (DP) and Direct-Active (DA) targeting.

## CONCLUSIONS

It is certain that nanorobotics if adopted for targeted interventions will have an impact on medical practice where the computer programmed with laws of physic being applied at the macroscale and the nanoscale will play a more important role, especially during the intervention. After the tumor has been identified, localized and characterized with the proper imaging modalities as well as gathering information about blood flow velocities and other physiological data, specific information will be entered in a special computer program. This program will be developed to assist the medical staff during the pre-operative phase prior to use the nanorobotic interventional platform. Based on the physiological data, the computer program will recommend one or a combination of specific nanorobots (*e.g.* TMMC of different sizes, concentration of active or passive MNP, *etc*; MTB, TMTB, *etc.*) from a library of available navigable entities. Once approved by the medical staff, the characteristics of the nanorobots with the intended volume to be injected will be entered automatically in the real-time navigation module. Based on that, a proposed trajectory and release site will appear on a computer display for final approval. Once approved, the coordinates will be automatically entered in a real-time navigation module.

At this moment, the patient under general anesthesia will be laid in the bore of the Magnetic Resonance Targeting (MRT) system (*i.e.* an upgraded MRI scanner capable of direct targeting interventions). Then, since MR-tracking of the nanorobots relies on their magnetic signatures, an off-resonance imaging sequence will first be performed to ensure that the volume of interest is free from any magnetic perturbations that could interfere with the tracking operation during the real-time navigation phase. Injections will follow prior to the real-time navigation initially using closed-loop trajectory control complemented with computer-based triggered release actions with the possibly of navigating TMTB deeper in the vasculature. In the future, the real-time process being performed will vary according to the type of interventions and various medical practices will be tested and adapted to take into considerations the potential offered by these new technologies.

# REFERENCES

[1]     http://www.davincisurgery.com/davinci-surgery/davinci-surgical-system/
[2]     http://www.givenimaging.com
[3]     http://genome.jgi-psf.org/magm1/magm1.home.html
[4]     Martel S, Mohammadi M, Felfoul O, *et al.* Flagellated magnetotactic bacteria as controlled MRI-trackable propulsion and steering systems for medical nanorobots operating in the human microvasculature, International Journal of Robotics Research IJRR 2009;28:571-582.
[5]     http://en.wikipedia.org/wiki/Robot
[6]     http://www.dailymail.co.uk/health/article-1037186/Te
[7]     http://www.vector-project.com/
[8]     Ergeneman O, Dogangil G, Kummer MP *et al.* A magnetically controlled wireless optical oxygen sensor for intraocular measurements. IEEE Sensors J 2008;8:29-37.
[9]     Martel S, Felfoul O, Mathieu JB *et al.* MRI-based nanorobotic platform for the control of magnetic nanoparticles and flagellated bacteria for target interventions in human capillaries. IJRR 2009 Special Issue on Medical Robotics;28:1169-1182.
[10]    Bell DJ, Leutenegger S, Hammar KM *et al.* Flagella-like propulsion for microrobots using a nanocoil and a rotating electromagnetic field. In Proc. of the IEEE Int. Conf. on Robotics and Automation (ICRA) 2007;1128-1133.
[11]    Mathieu JB, Martel S. Aggregation of magnetic microparticles in the context of targeted therapies actuated by a magnetic resonance imaging system. J App Phys 2009;106:44904-1 to 7.
[12]    Pouponneau P, Leroux JC, Martel S. Magnetic nanoparticles encapsulated into biodegradable microparticles steered with an upgraded magnetic resonance imaging system for tumor chemoembolization. Biomaterials 2009;30:6327-6332.
[13]    Wilson MW, Kerland RK, Fidelman NA *et al.* Hepatocelluler carcinoma: Regional therapy with a magnetic targeted carrier bound to Doxorubicin in a dual MR imaging/conventional angiography suite – Initial experience with four patients. Radiology 2004;230:287-293.
[14]    Martel S, Mathieu JB, Felfoul O *et al.* Automatic navigation of an untethered device in the artery of a living animal using a conventional clinical magnetic resonance imaging system. App Phys Lett 2007;90:114105.
[15]    Chanu A, Felfoul O, Beaudoin G *et al.* Adapting the software platform of MRI for the real-time navigation of endovascular untethered ferromagnetic devices. Mag Reso Med 2008;59:1287-1297.
[16]    Tamaz S, Chanu A, Mathieu JB *et al.* Real-time MRI-based control of a ferromagnetic core for endovascular navigation, IEEE Transac Biomed Engin 2008;55:1854-1863.
[17]    Felfoul O, Mathieu JB, Beaudoin G *et al.* MR-tracking based on magnetic signature selective excitation. IEEE Transac Med Imaging 2008;27:28-35.
[18]    Mathieu JB, Beaudoin G, Martel S. Method of propulsion of a ferromagnetic core in the cardiovascular system through magnetic gradients generated by an MRI system. IEEE Transac Biomed Engin 2006;53:292-299.
[19]    Mathieu JB, Martel S. *In vivo* validation of a propulsion method for untethered medical microrobots using a clinical magnetic resonance imaging system. In Proc. of the EEE/RSJ Int. Conf. on Intelligent Robots and Systems (IROS). San Diego, CA, USA, 2007
[20]    Fidleris V, Whitmore RL. Experimental determination of the wall effect for spheres falling axially in cylinder vessels. Br J Appl Phys 1961;12:490-494.
[21]    http://upload.wikimedia.org/wikipedia/commons/1/15/Flagellum_base_diagram_en.svg).
[22]    Berg HC, Brown DA. Chemotaxis in Escherichia coli analyzed by three-dimensional tracking. Nature 1972;239:500-504.
[23]    Ford RM, Phillips BR, Quinn JA *et al.* Measurement of bacterial random motility and chemotaxis coefficients. I. Stopped-flow diffusion chamber assay. Biotech Bioeng 1991;37:647-660.
[24]    Armitage JP. Bacterial motility and chemotaxis. Sci Progr 1992;76:451-477.
[25]    Frankel RB, Blakemore RP. Navigational compass in magnetic bacteria. J Magn Magn Materials 1980;15-18:1562-1564.
[26]    Denham C, Blakemore R, Frankel R. Bulk magnetic properties of magnetotactic bacteria. IEEE Trans Magne 1980;16:1006-1007.
[27]    Debarros H, Esquivel DMS, Farina M. Magnetotaxis. Sci Progr 1990;74:347-359.

[28]   Martel S, Tremblay C, Ngakeng S *et al.* Controlled manipulation and actuation of micro-objects with magnetotactic bacteria. App Phys Lett 2006;89:233804-6.

[29]   Lapointe J, Martel S. Thermoresponsive hydrogel with embedded magnetic nanoparticles for the implementation of shrinkable medical microrobots and for targeting and drug delivery applications. The 31st Annual International Conference of the IEEE Engineering in Medicine and Biology Society (EMBC'09), Minneapolis. USA, Sept. 2-6, 2009

[30]   Tabatabaei N, Martel S. The concentration effect of magnetic iron oxide nanoparticles on temperature change for hyperthermic drug release applications *via* AC magnetic field. The 5th International Conference on Microtechnologies in Medicine and Biology (MMB 2009) Conference, Quebec City, Canada, April 1-9, 2009

[31]   Hashizume H, Baluk P, Morikawa S, *et al.* Openings between defective endothelial cells explain tumor vessel leakiness. Am J Pathol 2000;156:1363-1380.

# CHAPTER 8

# Role of Nanotechnology in the Diagnosis and Treatment of Alzheimer's Disease

**Kristen M. Jaruszewski, Rajesh S. Omtri and Karunya K. Kandimalla**[*]

*Division of Basic Pharmaceutical Sciences, Florida A&M University, Tallahassee, FL 32307, USA*

**Abstract:** Nanotechnology has immense potential to revolutionize the treatment and diagnosis of neurodegenerative diseases such as the Alzheimer's disease (AD). AD is characterized by parenchymal amyloid plaques; intraneuronal tangles; substantial neuronal loss in the cortex and hippocampus; and significant cognitive decline. Treatment in the early stages of the disease is critical to halt or even reverse the neurodegeneration associated with AD. However, none of the diagnostic methods existing today can provide a definitive pre-mortem diagnosis of AD; currently available AD treatments can only offer symptomatic relief but do not address the underlying pathology. Nanotechnology is enhancing the sensitivity and specificity of magnetic resonance imaging and positron emission tomography contrast agents that can detect various pathological hallmarks of AD such as amyloid plaques and neurofibrillary tangles. In addition, nanotechnology enables the targeted delivery of novel therapeutic compounds that are being developed for AD treatment; enhance their efficacy; and reduce systemic toxicity. Some of these notable advances in AD diagnosis and treatment propelled by the nanotechnology are reviewed in this chapter.

**Keywords:** Alzheimer's disease; amyloid plaques; neurofibrillary trangles; contrast imaging; smart nanovehicles; biomarkers; treatment; pharmacology; nanotechnology; genetics.

## 1. ALZHEIMER'S DISEASE

Alzheimer's disease (AD) is a devastating neurodegenerative disease named after Dr. Alois Alzheimer who described it for the first time in a mentally ill German woman [1]. Currently, 5 million Americans and 26 million people world-wide are afflicted with AD [2]. According to some estimates the number of AD patients are expected to increase to 105 million by 2050, if no advances in the early diagnosis and treatment of the disease are made [2]. Risk of being diagnosed with AD increases with age; an individual between ages 65 and 74 years may have a 10% chance of developing AD, but the risk increases substantially to 47% at 85 years or older [3]. In the early stages of the disease, an AD patient may experience lapses in short-term memory and behavioral disturbances, which are not readily distinguishable from the normal signs of aging; hence it is very difficult to diagnose AD at this stage [4]. However, in AD patients these symptoms worsen rapidly leading to drastic changes in cognitive and bodily functions.

### 1.1 Genetic Predisposition to AD

Unlike the age-dependent sporadic AD described above, the familial AD, which accounts for a small percentage (<5%) of total AD cases, is genetically inherited. Individuals with mutations on genes for amyloid precursor protein (APP), presenilin 1 (PSEN1), and presenilin (PSEN2) can inherit familial AD in an autosomal dominant manner [5-6] and suffers from AD at an early age of 55-60 years [7]. Although, specific genes responsible for sporadic AD have not yet been identified, ApoE gene on chromosome 19 that codes for apolipoprotein E (ApoE) appears to increase as individual's risk of developing sporadic AD. The ApoE modulates the transport of triglycerides, lipoproteins, and phospholipids and is believed to play a role in nerve regeneration and synaptic remodeling [8]. In humans, the ApoE exists as three isoforms APOE ε2, APOE ε3, and APOE ε4 [9-10], out of which the ApoE e4 is associated with AD [11]. Numerous studies attest to the fact that the ApoE e4 allele is 3 to 4 times more common in sporadic AD patients than in individuals who are not demented [11]. Nevertheless, ApoE e4 is not a deterministic cause of AD, only a risk factor [12].

**\*Address correspondence to Karunya K. Kandimalla:** Division of Basic Pharmaceutical Sciences, Florida A&M University, Tallahassee, FL 32307, USA; Tel: 850 599 3581; E-mail: karunya.kandimalla@famu.edu

**Mark Slevin (Ed)**

## 2. AD PATHOPHYSIOLOGY

Loss of neurons in the cortex and hippocampus is the main neuropathologic feature underlying the AD symptoms. Radiological examination of AD brain may indicate brain atrophy associated with narrowed gyri and enlarged ventricles and sulci [13]. However, these features are not specific to AD and may also be present with other dementias. Histological examination of post-mortem AD brain may reveal the presence of amyloid plaques, neurofibrillary tangles (NFTs), and vascular amyloid deposits. The amyloid plaques are extracellular deposits of amyloid beta (Aβ) proteins in the brain parenchyma and NFTs are formed in the neurons due to abnormal rearrangement of microtubule-associated tau proteins (Fig. **1**). In addition to being deposited as amyloid plaques, the Aβ proteins are also deposited in the walls of medium-and small leptomenengial and cortical arteries to cause CAA (Fig. **2** [14]), which may lead to vascular dementia, massive lobar hemorrhages and stroke [15]. Approximately 10 to 40 % of elderly subjects are diagnosed with CAA. But in AD patients, the incidence of CAA increases to 80% [16]. The presence of amyloid deposits and NFTs in the brain is considered as the primary pathological hallmark of AD.+

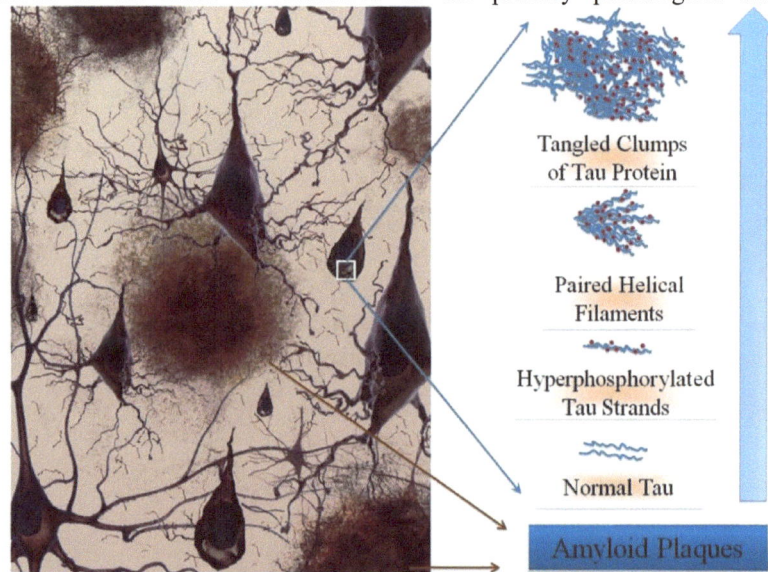

**Figure 1: Hallmarks of Alzheimer's disease.** Parenchymal amyloid plaques and intra-neuronal tangles of hyperphosphorylated tau protein are considered as the primary hallmarks of Alzheimer's disease. Modified from: http://www.nia.nih.gov/Alzheimers/Publications/Unraveling/Part2/hallmarks.htm

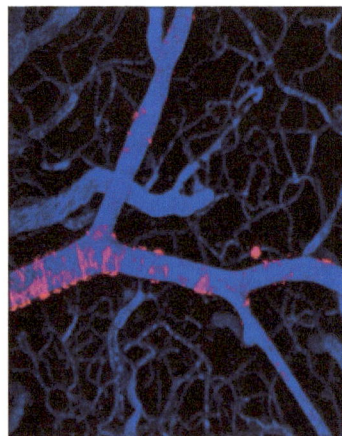

**Figure 2: Cerebral amyloid angiopathy (CAA) in a living AD transgenic mouse (Tg2576) imaged with multiphoton microscopy.** Blood vessels and vascular Aβ deposits are identified by systemically administered Texas Red dextran (blue pseudocolor) and methoxy-XO4 (red), respectively (14).

## 2.1. Role of Aβ Proteins and NFTs in Neurodegeneration

The Aβ proteins 40 and 42, which are the primary constituents of the parenchymal amyloid plaques and cerbrovascular amyloid deposits, are derived from a large transmembrane glycoprotein known as the amyloid precursor protein (APP) following successive cleavages by β- and γ-secretases [17]. In healthy subjects, the APP could also be processed by α- and γ-secretases that results in a shorter non-pathological p3 fragment [18]. The patho-physiological factors that trigger β- and γ -secretase processing of APP versus the α -secretase processing is not known.

Despite being small proteins, Aβ40 even more so Aβ42 have great propensity to self aggregate. They exist in various aggregation states *in vivo* ranging from low molecular weight oligomers, protofibrils, and mature fibrils that eventually come together as parenchymal plaques or cerebrovascular amyloid deposits (Fig. **3**). Recent research indicates that soluble low molecular weight Aβ aggregates (sAβ) are more toxic to neurons and the BBB endothelial cells than the insoluble fibrils [19-20]. The better studied of sAβ oligomers are: Aβ-derived diffusible ligands (ADDLs) [21], oligomers composed of 15–20 monomers (AβOs) [22], protofibrils [23], and dodecameric oligomers (Aβ*56) [24]. The Aβ oligomers that are smaller than 50 kD are generally considered as ADDLs, whereas the other oligomers usually have a molecular mass of 50-150 kD. In rodents, the sAβ oligomers were shown to be neurotoxic at low concentrations, inhibit long-term potentiation, and lead to cognitive dysfunction [21, 24, 25]. Most importantly, the brain level of soluble Aβ species is believed to correlate with the loss of cognitive function better than the amyloid plaque burden.

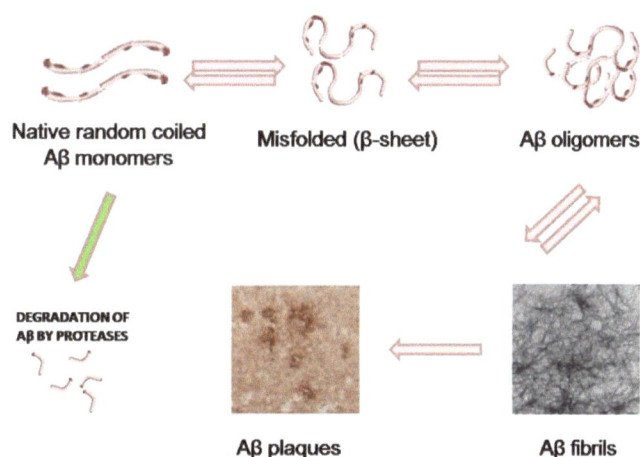

**Native random coiled Aβ monomers**    **Misfolded (β-sheet)**    **Aβ oligomers**

**DEGRADATION OF Aβ BY PROTEASES**

**Aβ plaques**    **Aβ fibrils**

**Figure 3: Aβ aggregation.** Aβ monomers that normally exist as random coils are misfolded to form β-sheet structures. Soluble low-molecular weight aggregates and insoluble amyloid fibrils are generated from the misfolded Aβ and subsequently deposited in AD brain as amyloid plaques and cerebrovascular deposits.

While amyloid plaques and NFTs actually define AD, they do not fully represent the disease process. It is currently believed that the intraneuronal accumulation of Aβ proteins is the first step in a fatal cascade of events leading to neurodegeneration in AD (reviewed by Wirths [26] and Laferla [27] ). Upon accumulation, Aβ has been reported to disrupt the normal functioning of neurons resulting in significant cellular dysfunction leading to apoptosis [28] and oxidative injury [29], even before the formation of senile plaques and neurofibrillary tangles. Over-production of Aβ proteins in AD subjects and alteration of Aβ disposition in various body compartments during the disease progression may increase the exposure of neurovascular unit to toxic Aβ proteins. Chronic exposure to various toxic forms of Aβ protein may disrupt the BBB and compromise its ability to clear Aβ proteins [30]. As a result, Aβ accumulates in the brain parenchyma where it is taken up by cortical and hippocampal neurons. Therefore, most of the AD diagnostic procedures are designed to detect and quantify the accumulation of Aβ proteins in various physiological compartments, whereas the novel treatment methods are usually aimed at decreasing the production of Aβ proteins and/or increase their elimination from the brain [31].

## 2.2 Patterns of Aβ Protein Production and Clearance

The Aβ proteins are generated in the peripheral tissues as well as the brain. It has been reported that Aβ levels in the brain tissue, CSF, and plasma are in equilibrium, which is disturbed during AD progression (Fig. **4**) [32, 33]. The Aβ proteins circulating in the plasma are regulated by renal clearance and hepatic elimination. In addition, the BBB is believed to regulate Aβ concentrations in the plasma and brain compartments *via* the RAGE (receptor for advanced glycation end products) receptor that transports Aβ proteins in the luminal to abluminal direction [34] and LRP1 (low density lipoprotein receptor-related protein 1) receptor that transports them in the abluminal to luminal direction [35]. In the brain parenchyma Aβ proteins might either aggregate to form fibrils or subjected to enzymatic degradation by various proteases such as insulin degrading enzyme (IDE) and neprilysin (NEP). Heavy metals like Iron (Fe), Zinc (Zn) and Copper (Cu) accelerate the aggregation of Aβ proteins [36, 37]. These Aβ aggregates cause neuronal toxicity and also provide nidus for the formation amyloid plaques [38].

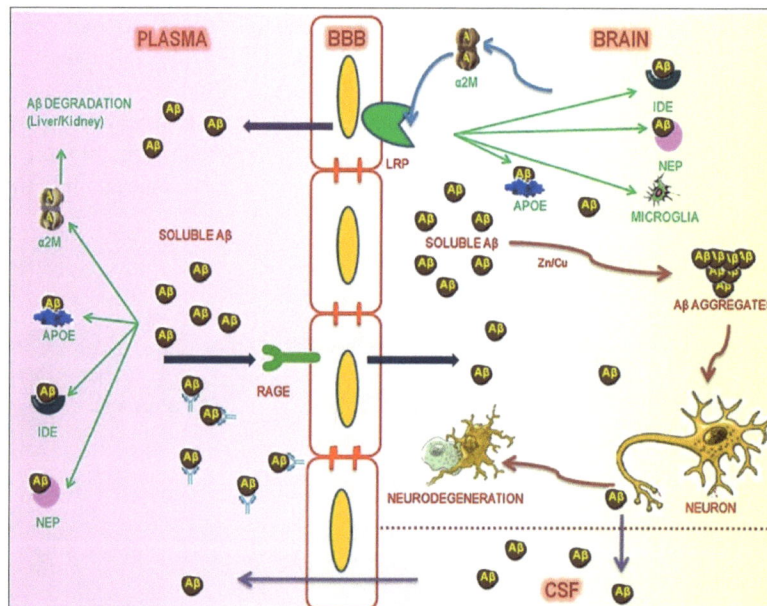

**Figure 4: Aβ biodistribution.** The Aβ proteins are produced in the brain as well as in the peripheral tissues. In a healthy individual, the plasma and brain Aβ levels are maintained at equilibrium. After binding to chaperone molecules such as α-2-macroglobulin (α2M) and apolipoprotein E (apoE), Aβ proteins in the plasma are either cleared by liver and kidney or transcytosed across the BBB via the RAGE receptor. Aβ produced in the brain could: aggregate in the presence of heavy metals like zinc (Zn) and copper (Cu) to form amyloid plaques; be internalized by neurons and cause neurodegeneration; degraded by neprilysin (NEP) and insulin degrading enzyme (IDE); or effluxed into plasma (mostly) by LRP1 receptor (34).

## 3. DIAGNOSIS OF AD AND CAA

Early diagnosis of AD is of paramount importance, because therapeutic intervention in the beginning stages of the diseases could limit the neuronal loss and provide opportunities to halt or even reverse AD progression. Pre-mortem clinical indications such as behavioral changes and neuropsychological tests are often employed to assess the extent of cognitive impairment and severity of the disease. However, these methods could be subjective, imprecise, and may not provide any assessment of AD pathology. A definitive AD diagnosis is currently possible only post-mortem by verifying the presence of histopathological hallmarks such as extracellular senile plaques and neurofibrillary tangles. Sensitive imaging technologies and specific contrast agents capable of detecting these histopathological hallmarks in live AD patient; development of quantitative magnetic resonance (MR) techniques that measure the anatomic, biochemical, microstructural, functional, and blood-flow changes indicative of AD pathophysiology [39]; and ability to assay biomarkers such as ADDLs and neurofibrillary tangles in various physiological compartments may be critical for the early diagnosis of AD and to monitor the disease progression.

### 3.1 Imaging Amyloid Plaques

In the recent years, substantial effort has been focused on the development of imaging techniques such as magnetic resonance imaging (MRI) and positron emission tomography (PET) to visualize amyloid plaques in AD patients. Jack *et al.* have demonstrated that 35 micron diameter amyloid plaques can be detected without any contrast agent under high magnetic field strength (9.4 Tesla) [40]. But in clinical practice much lower strength magnets are used (3 or 7 Tesla). Moreover, diffuse plaques formed in the early stages of AD are much smaller than 35 microns. Therefore, MRI in conjunction with a contrast agent is required to resolve individual plaques and also to differentiate plaques from other interfering structures such as blood vessels, myelinated fibers, and intracranial structures [41].

#### 3.1.1 MRI Contrast Agents

Of the MRI contrast agents that are currently being developed for imaging amyloid plaques, the most notable is the amyloid protein itself, mostly $A\beta40$ [41-45]. The $^{125}I$ labeled $A\beta40$ ($^{125}I$-$A\beta40$) was reported to have high binding affinity to the amyloid plaques in human and double transgenic AD mouse (APP,PS1) brain slices *in vitro*. However, the plaque targeting ability of $^{125}I$-$A\beta40$ after intravenous injection in APP,PS1 mice was low, most likely due to lower permeation at the BBB [42]. Therefore, the $A\beta40$ protein was conjugated to a naturally occurring polyamine such as putrescein to increase its permeability across the BBB. In addition, gadolinium diethylenetriaminepentaacetic acid (Gd-DTPA), which would be necessary to provide MR contrast enhancement for the labeled plaques, was covalently attached to $A\beta40$. When injected in APP,PS1 animals Gd-DTPA conjugated $A\beta40$ targeted amyloid plaques and provided T2 contrast that corresponded with the signal on T1 spin echo [42].

#### 3.1.2 PET Contrast Agents

In addition to MRI, the PET coupled with radiolabeled agents that specifically recognize and bind to amyloid plaques have been developed. The most promising of these agents, $^{11}C$ labeled 6-hydroxy-2-(4'-N-[$^{11}C$]methyl-aminophenyl-1,3-benzothiazole, also known as Pittsburg compound B ($^{11}C$-PIB), is currently being evaluated in multi-center clinical trials. The $^{11}C$-PIB was shown to bind a variety of amyloid plaques including diffuse plaques and dense core plaques [46]. In addition, $^{11}C$-PIB was shown to bind cerebrovascualr amyloid deposits [46]. PET scan following the administration of $^{11}C$-PIB has shown significantly higher $^{11}C$-PIB retention in the gray matter of AD patients than in healthy controls (HCs) (Fig. **5** [47]). It has been shown that the extent of $^{11}C$-

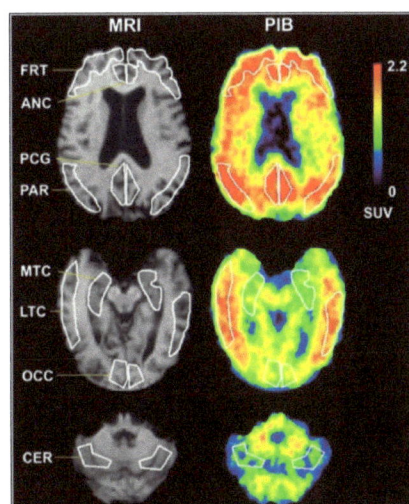

**Figure 5: Examples of magnetic resonance (MR) images (left) and Pittsburgh compound-B ($^{11}C$-PIB) positron emission tomography (PET) images of Alzheimer's disease brain (right).** The images were acquired from a 77-year-old male AD subject with moderate levels of $^{11}C$-PIB retention. Shown in white are the regions-of-interest examined in the study (FRT: frontal, PCG: posterior cingulated gyrus, PAR: parietal, MTC: mesial temporal, LTC: lateral temporal, OCC: occipital, and CER: cerebellum (reference region). The PET images were summed over 40 to 60 min after injection and normalized to injected dose and body mass [47].

PIB accumulation strongly correlates to the clinical progression of the disease [48, 49]. However, the limitation of [11]C-PIB contrast agent is that it binds to the white matter of AD and HC subjects in a nonspecific and non-saturable manner [50]. Moreover, PET imaging does not have high enough resolution to image individual plaques. Usually PET imaging offers a reliable AD diagnosis when combined with the assays of other biomarkers such as the estimation of Aβ protein concentrations in the plasma and/or Cerebrospinal Fluid (CSF).

### 3.1.3 Nanotechnology in the Detection of Parenchymal Amyloid Plaques

The challenge of designing a nanoprobe to image parenchymal amyloid plaques could arise from the difficulty of the probe to permeate the BBB and the diffusional resistance it may encounter while navigating through a highly tortuous brain parenchymal tissue. The endothelium of cerebral microvasculature with well formed tight junctions constitutes the blood brain barrier (BBB), which restricts the CNS delivery of molecules above 600 daltons. The uptake of nanoparticles across the BBB is usually mediated by phagocytosis, carrier-mediated endocytosis, or adsorptive endocytosis. The carrier-mediated endocytosis is more efficient in delivering cargoes across the BBB than phagocytosis. Therefore, many attempts have been made to conjugate a cell surface receptor/transporter ligand on the surface of the nanoparticle to promote its uptake by the BBB endothelial cells *via* carrier-mediated endocytosis (reviewed by Bareford and Swaan [51]). Aktas *et al.* [52] have demonstrated that nanoparticles bearing OX26 monoclonal antibody that targets the transferrin receptor, trigger transcytosis of nanoparticles across the BBB [53]. Alternatively, we developed methods to promote electrostatic interactions of nanoparticles with the endothelial cell surface so that they could be internalized by adsorptive endocytosis. When conjugated with polyamines that increase positive charge density on the surface, a nanoparticle could adsorp to the negatively charged endothelial cell surface and subsequently be internalized *via* adsorptive endocytosis. Some of these methods are employed to improve the BBB permeability of a nanoprobe that can target amyloid plaques and provide contrast enhancement for MR imaging.

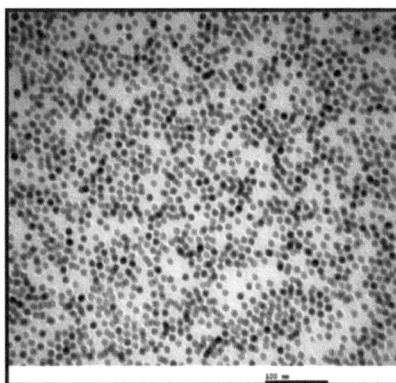

**Figure 6:** Transmission electron micrograph of monocrystalline iron oxide nanoparticles (MIONS).

Monocrystalline Iron Oxide Nanoparticles (MIONs) (Fig. **6**) have been previously employed to image amyloid plaques in AD transgenic animals [45]. Owing to their high magnetic moments, MIONs require lower tissue concentrations for similar contrast enhancement than a conventional MRI contrast agent such as gadolinium diethylenetriaminepentaacetic acid (Gd-DTPA). Iron oxides such as magnetite ($Fe_3O_4$) [54] or maghemite (g-$Fe_2O_3$) [55] are the most preferred MR contrast agents due to their stability in aqueous solutions and their high saturation magnetization at body temperature and at low field. MIONs are usually 10-20 nm in size and have a high propensity to aggregate due to their surface hydrophobicity. Hence, they are usually coated with water-soluble polymers such as dextran (commonly used as a plasma expander) and chitosan; surfactants such as sodium oleate and dodecylamine; and phospholipids (reviewed by Gupta and Gupta [56]). Surface coating with dextran is popular because of its high affinity to iron oxide [57]. Inspite their small size, delivery of MIONs across the BBB is very challenging. Wadghiri *et al.* have successfully used osmotic methods, often employed to deliver chemotherapeutic agents to the brain tissue, to deliver Aβ40 conjugated MIONs to the amyloid plaques in the brain parenchyma. In their experiments, mannitol was used as an osmotic agent to transiently open the BBB. Intra-arterial administration of Aβ40 conjugated

MIONs with mannitol enabled their accumulation on the amyloid plaques and provide contrast for MR imaging (Fig. **7** [45]).

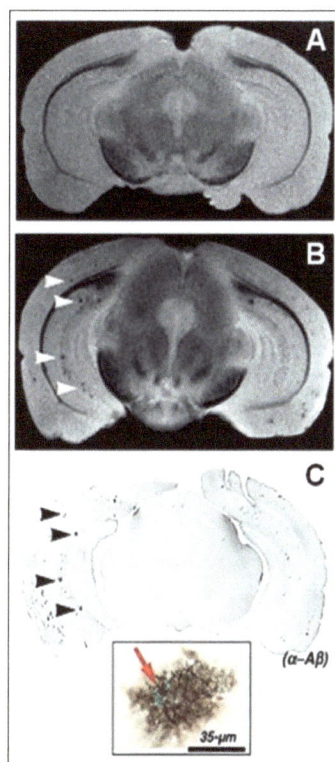

**Figure 7:** Aβ plaques were detected with *ex vivo* µMRI after injection of Gd-DTPAAβ40 with mannitol. *Ex vivo* T2-weighted (SE coronal µMR images show 6-monthold control (a) and AD transgenic (b) mouse brains. Both brains were extracted and prepared for imaging 6 hr after carotid injection of Gd-DTPA-Aβ40 with 15% mannitol. Note the obvious matching of many larger plaques (arrowheads) between µMRI (b) and immunohistochemistry (c). FromWadghiri *et al.,* 2003 [45].

Fluorescence emitting semiconductor nanocrystals, popularly known as Quantum dots (Qdots) could also serve as contrast agents for imaging amyloid plaques. The Qdots are usually around 10 nm in size and have novel optical and electrical properties. By changing the size and composition of Qdots, their emission wavelengths can be tuned from visible to infrared regions. The Qdots have large absorption coefficients across a wide spectral range, demonstrate high levels of brightness, and maintain photo stability [21]. They can be targeted to various pathophysiological structures by attaching antibodies and peptides and then visualized using near-infrared (NIR) optical imaging [36]. The downside of quantum dots include: poor stability and aqueous dispersibility; lack of BBB permeability; and toxicity. The aqueous dispersibility of Qdots could be improved by coating them with amphiphilic polymers containing a hydrophobic segment (hydrocarbon chain) and a hydrophilic group such as polyethylene glycol (Reviewed by Hezinger *et al.* [58]). Currently, our laboratory is focusing on targeting Qdots to amyloid plaques in the AD brain by conjugating them with novel anti-amyloid antibodies such as IgG4.1.

In addition, our lab has developed a polymeric smart nanovehicle capable of carrying MIONs Qdots or Gd-DTPA across the BBB and target cerebrovascular amyloid deposits (Fig. **8**) [59]. The basic design of SNV comprises of a drug and/or a diagnostic agent entrapped in a Chitosan Polymeric Core (CPC) with appropriate size and surface characteristics. The chitosan was selected to prepare the polymeric core because it is biocompatible and does not cause allergic reactions. Moreover, chitosan is bioadhesive and is known to improve drug absorption at cellular barriers mostly due to its high positive charge density. A cationized F(ab')₂ fragment of IgG4.1 was conjugated on the surface of the CPC to enhance its uptake at the BBB. For maximal retention in the vascular wall the SNVs were maintained around 200-240 nm.

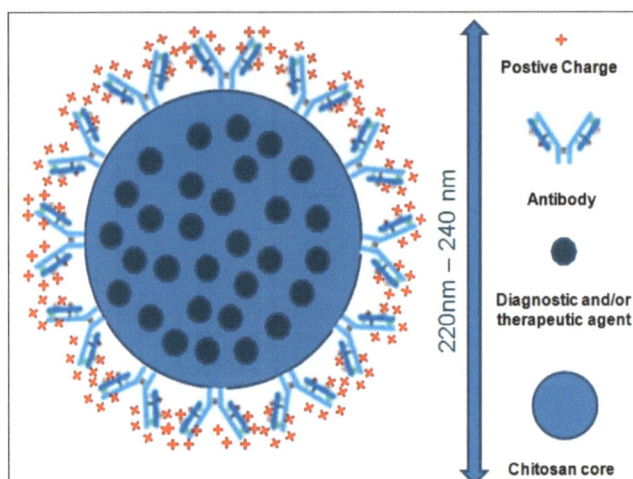

**Figure 8:** Smart nanovehicle carrying diagnostic and therapeutic agents.

## 3.2 Biomarkers

Extensive research is being conducted to identify biomarkers that reveal AD pathology and disease progression. It is beyond the scope of this article to extensively summarize and assess all of the potential AD biomarkers. Instead, the use of Aβ and tau proteins as biomarkers will be discussed because they are structurally and pathophysiologically related to amyloid plaques and neurofibrillary tangles. While the levels of soluble Aβ and tau increase in the bloodstream as well as in the CSF with AD progression, the CSF levels present a more reliable diagnostic criterion than the plasma levels. In AD patients, a decreased concentration of CSF Aβ42 (*i.e.*, <500 ng/L) and an increased concentration of total tau protein (>350 ng/L) or hyperphosphorylated tau (>85 ng/L) was observed [60]. The assessment of both total tau and Aβ42 concentrations in CSF was shown to yield a sensitivity of 81% to 94% and a specificity of 79% to 95% for differentiating between AD patients and controls [61]. Hence the detection of Aβ and tau levels in the CSF forms the basis for the development of AD diagnostic tests.

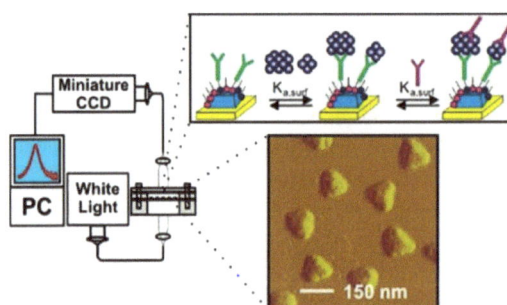

**Figure 9: Design and experimental setup of localized surface plasmon resonance (LSPR) biosensor for the detection of ADDLs.** Transmission UV-VIS spectroscopy is used to monitor the optical properties (LSPR) of Ag nanoparticles. The schematic illustration displays the sandwich assay and surface chemistry of the LSPR nanosensor. First, surface-confined Ag nanoparticles (see atomic force microscopy inset) are synthesized using nanosphere lithography on mica substrates. Next, a self-assembled monolayer consisting of a mixed monolayer of 1-octanethiol and 11-Mercaptoundecanoic acid passivates the nanoparticles for nonspecific binding and activates the nanoparticles for the attachment of the first anti-ADDL antibody, respectively. The first anti-ADDL antibody is covalently attached to the nanoparticles *via* incubation in 100 mM 1-Ethyl-3-[3-dimethylaminopropyl]carbodiimide hydrochloride/100 nM anti-ADDL antibody solution for 1 h. Samples are then incubated in varying concentrations of ADDLs for 30 min. Finally, to enhance the LSPR shift response of the ADDLs, the samples are incubated in a 100 nM anti-ADDL solution for an additional 30 min [64].

## 3.3 Nanotechnological Methods to Detect AD Biomarkers

High levels of amyloid derived diffusible ligands (ADDLs) are found in the brain of AD patient. ADDLs are thought to be present in CSF and possibly the blood, allowing for a reliable and noninvasive diagnosis tool. In light of the invasiveness involved in sampling CSF, recent research has been focused on developing a highly sensitive, accurate technique which measures the concentration of ADDLs using localized surface plasmon resonance (LSPR) nanosensor coupled with anti-amyloid antibodies [62]. This technique is very similar to sandwich enzyme linked immune-sorbent assay (ELISA). Briefly, Silver (Ag) nanoparticles immobilized on a mica substrate were conjugated with antibodies that specifically bind to ADDLs. Physiological fluids, plasma or CSF, containing ADDLs (antigen) were flown on the surface for 30 minutes to allow the antibody on the surface of the Ag nanoparticles to capture the ADDLs. Finally, a polyclonal anti-ADDL antibody was applied (schematic 1) to detect the ADDLs captured from the plasma or CSF. It was shown that there is an exceptionally high affinity (1fM-10 pM) between the ADDL molecules captured on the surface and anti-amyloid antibody. However, this method requires a high concentration <10 pM ADDL of antigen, which is prevalent in the advanced stages of the disease, thus limiting its applicability as an early diagnostic tool.

Another procedure proposed to detect ADDLs is the bio-bar code assay developed by the Mirkin and associates [63]. The bio-bar code assay was claimed to be effective in detecting ADDLs at the attomolar range [64]. Currently, it is being employed to test mad cow disease and HIV. The bio-bar assay utilizes antibodies, magnetic microparticles, gold nanoparticles, and DNA. An ADDL specific monoclonal antibody (20C2) is immobilized on the surface of magnetic nanoparticles. While the gold nanoparticles are functionalized with thiolated DNA and anti-ADDL polyclonal (M90) antibody. Then the complementary bar-code DNA is added and hybridized to the gold nanoparticle. This process involves two significant binding events. The first event is related to the recognition and attachment of ADDL to the monoclonal anti-ADDL antibody on the magnetic nanoparticle. This complex is magnetically separated from any unbound molecules and repeatedly washed. The second binding event involves the recognition and attachment of the polyclonal anti-ADDL antibody (linked to the gold nanoparticles) to the ADDLs captured by the magnetic microparticle. The ADDLs, gold nanoparticles, and magnetic microparticles are again isolated through magnetic separation and washed repeatedly. Finally, with the addition of water, the complementary DNA on the nanoparticle dehybridizes and is released (Fig. **10**). When the ADDL concentration is increased, there is a corresponding increase in the formation of ADDL-magnetic microparticle-gold nanoparticle complexes. These complexes trigger increased release of DNA and a corresponding to amplification of the signal. The major limitation of this method is that it is qualitative and does not provide information on the amount of ADDLs present in the physiological matrices.

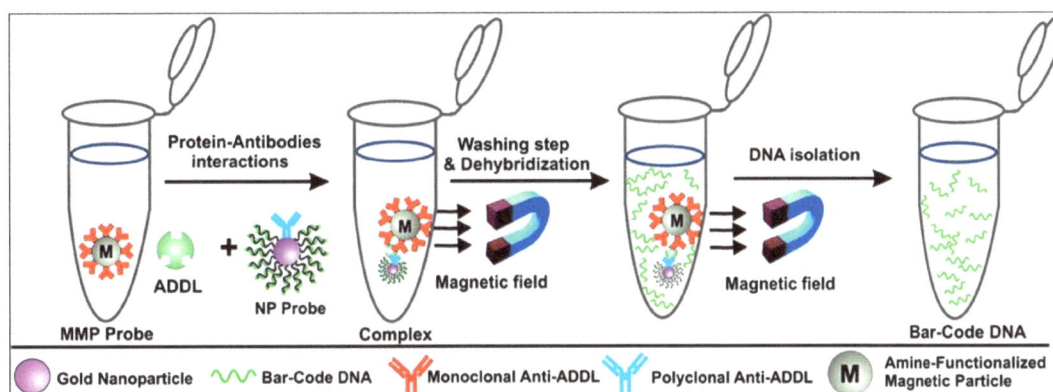

**Figure 10: The bio-barcode amplification assay.** The assay uses magnetic microparticles (MMPs) functionalized with monoclonal antibodies that recognize and bind ADDLs. The ADDLs are then sandwiched with gold nanoparticles, modified with double-stranded DNA and anti-ADDL polyclonal antibodies. After repeated washing while using a magnet to immobilize the MMPs, a dehybridization step releases hundreds of barcode DNA strands for each antigen-binding event. From Georganopoulou *et al.,* 2005 [66].

## 4. NANOTECHNOLOGY IN THE TREATMENT OF AD AND CAA

While a permanent cure for AD is not yet a reality, some currently available drugs offer symptomatic relief and may help alleviate cognitive and behavioral deficits associated with AD. Recent advances in biotechnology coupled with better understanding of AD pathogenesis have facilitated the development of new classes of therapeutic compounds that specifically target AD pathology. However, many of these compounds degrade in the body quickly; permeate poorly across biological barriers such as the BBB; and may cause systemic toxicity. Nanoparticles allow for delivering such therapeutic agents to the target site and limit systemic toxicity. In addition, the nanoparticles also protect the molecules from enzymatic degradation in the physiological compartments. Various types of nanoparticles that are useful in AD treatment include: a) polymeric nanoparticles, b) inorganic nanoparticles, c) liposomes, and d) solid lipid nanoparticles [65].

a) Polymeric nanoparticles are usually made of a nanocore composed of biodegradable and biocompatible polymers such as chitosan, polylactic-co-glycolic acid (PLGA), polyglycolic acid (PGA), polyalkylcyanoacrylate, and poly methylmethacrylate. A therapeutic and/or a diagnostic agent is usually entrapped in the polymeric nanocore.

b) Inorganic nanoparticles are traditionally made from silica, alumina, metals (*i.e.* germanium or gemanium silicon), metal oxides (*i.e.* manganese oxide, iron oxide, germanium oxide), and metal sulfides (*i.e.* cadmium selenide). These nanoparticles can be surface modified with metal chelators that are known to inhibit the aggregation of Aβ proteins [66].

c) Liposomes are composed of a phospholipid bilayer surrounding an aqueous core and range in size from 10 nm up to 1 micron. Liposomes are capable of encapsulating water-soluble compounds within the hydrophilic core and lipophilic compounds within the phospholipid bilayer. However, the bioavailability of very hydrophilic drugs in liposomes is low due to their lower encapsulation efficiency and also lower release rate from the liposome core [67].

d) Solid lipid nanoparticles (SLNs) are submicron sized nanoparticles that are more stable than liposomes. The SLNs consist of a rigid lipid nanocore made of triglycerides, fatty acids, and waxes coated with a monolayer of phospholipids. Lipophilic drugs and/or diagnostic agents are entrapped in the nanocore.

Often the surface of these nanoparticles is functionalized to improve aqueous dispersibility, bioadhesion, plasma residence time and uptake at the cellular barriers. The dispersibility of nanoparticles is enhanced by optimal charge density on the nanoparticle surface. An increase in the positive charge density would also enhance the bioadhesion of the nanoparticles to the negatively charged BBB endothelial cells, so that they are eventually internalized *via* adsorptive endocytosis [68]. The plasma residence time of nanoparticles is increased by covalently linking polyethylene glycol (PEG) to the nanoparticle surface. The PEG works by forming a hydrophilic coating around the nanoparticle so that its clearance by the reticuloendothelial system (a system of phagocytic cells present in liver, spleen, kidney, *etc.*) is significantly reduced (Reviewed by Alexis *et al.* [69]). To enhance the uptake of nanoparticles at the BBB, they are conjugated with ligands such as transferrin, low density lipoprotein, and insulin that are transcytosed across the BBB *via* receptor-mediated endocytosis [70-72]. Ulbrich *et al.* have demonstrated that human serum albumin nanoparticles covalently conjugated with transferrin or anti-transferrin antibodies (OX26 or R17217) successfully delivered loperamide (cannot corss the BBB) to the brain after intravenous injection [73]. Transferrin conjugated liposomes showed specific targeting to the brain in post-ischemic rats [74]. Likewise, 83-14 Mab (human insulin receptor antibody) conjugated to Pegylated liposomes (liposomes conjugated with polyethylene glycol) showed successful gene delivery to the brain [75].

## 5. CONVENTIONAL AD DRUGS

Acetylcholinesterase inhibitors such as tacrine, donepezil, rivastigmine, and galantamine are currently being used to slow the progression of mild to moderate AD. Nanoparticles could enhance the delivery of these drugs to achieve better efficacy/toxicity profiles. Rivastigmine encapsulated in polysorbate 80 coated poly-n-butylcyanoacrylate nanoparticles demonstrated a four fold increase in the brain uptake compared to the free drug [76]. Kreuter *et al.* hypothesized that the nanoparticles coated with polysorbate 80 adsorb apolipoprotein E (apoE) from the plasma and mimic LDL particles [77]. As a consequence they are endocytosed at the BBB *via* the LDL receptor. Moreover, polysorbate 80 coating increases the plasma residence time of nanoparticles by preventing opsonization and subsequent clearance by the reticuloendothelial system. Similarly, the distribution of polysorbate 80 coated and tacrine entrapped chitosan nanoparticles were found to be much lower to the liver and spleen, organs comprising of reticuloendothelial system, compared to the uncoated nanoparticles [78].

## 6. NOVEL DRUGS TO TREAT AD AND CAA

Despite the ability of nanoparticles to deliver conventional AD drugs to the brain, the drugs themselves are only marginally effective in treating AD. Therefore, many academic research groups and various pharmaceutical companies have invested substantial resources towards developing treatment strategies that target pathophysiological mechanisms of AD. Most of those strategies fall under the following categories:

### 6.1. Treatments to Reduce Aβ Production

The production of toxic Aβ proteins, generated by the processing of APP by β- and γ-secretases, is upregulated in AD (reviewed by Kim and Tsai [79]). This Aβ overproduction can be controlled by inhibiting the activity of β- and γ-secretases through pharmacological or genetic approaches.

#### *6.1.1. Pharmacological Approaches*

These agents include the inhibitors to β- and γ-secretase activity that generate Aβ40/Aβ42 by cleaving APP at C-711/713 amino acids and N-671/673 amino acids, respectively [80]. Both β- and γ-secretase inhibitors are not effective against already formed Aβ40 and Aβ42, however, they offer a promising alternative to preventing the further formation of Aβ40 and Aβ42. The efficacy and toxicity of many secretase inhibitors are currently being investigated in the pre-clinical and clinical trials. The γ-secretase inhibitors such as BMS-289948 and BMS-299897 were shown to decrease Aβ protein concentrations in both brain and plasma of AD (APP-YAC) transgenic mice when administered intravenously [81]. Similarly, LY-450139, a γ-secretase inhibitor in phase III clinical trials has been shown to reduce Aβ concentration in blood and cerebrospinal fluid in mild to moderate AD patients [82]. In addition to inhibiting γ-secretase activity, many γ-secretase inhibitors also inhibit the cleavage of Notch, a transmembrane receptor involved in regulating cell-fate decisions, and cause abnormalities in the gastrointestinal tract, thymus and spleen. Therefore, more emphasis is placed on discovering β-secretase inhibitors or specific γ-secretase inhibitors that do not inhibit Notch cleavage.

Another approach is to activate the functioning of α-secretase that cleaves APP into non-amyloidogenic fragments smaller than Aβ 40 or 42. Triglycerides containing Polyunsaturated Fatty Acids (PUFAs) that promote membrane integrity was reported to increase α-secretase activity. For instance, 1,3-caprylol-2-arachidonoyl glycerol was shown to increase α-secretase activity by 18% and was shown to reduce Aβ40 secretion in glioblastoma cells [83].

#### *6.1.2. Genetic Approaches*

Miller *et al.* have claimed that small interfering RNAs (siRNAs) could silence genes encoding APP and tau in cell culture models [84]. The siRNAs could also inhibit β- and γ-secretases that generate Aβ proteins. However, siRNAs are very delicate and can get easily degraded in the physiological environment (reviewed by Akhtar and Benter, 2007 [85]). To maintain efficacy *in vivo*, the siRNAs must be delivered intact to the appropriate neurons. The delivery of siRNAs to neurons in the brain can be achieved by encapsulating them in nanoparticles that can cross the BBB [86] or by expressing them in viral vectors. Singer *et al.* have demonstrated that silencing β-secretase activity using siRNAs delivered *via* lentiviral vectors reduced amyloid production in AD transgenic mice that express APP 'Swedish' mutation [87]. Translational blocking of APP-mRNA at the β-secretase-site of APP using antisense oligodeoxynucleotides reduced cerebral amyloid and acetyl cholinesterase activity in TG2576 AD transgenic mice [88].

### 6.2 Treatments to Reduce the Aggregation of Aβ Proteins

Low molecular weight oligomers of Aβ proteins are believed to be more toxic to neurons and BBB endothelial cells than the monomers or large fibrils. Therefore, methods to prevent or reduce Aβ aggregation are expected to reduce AD severity.

### 6.2.1. Metal Chelators

Metals like iron, copper or zinc accelerate Aβ aggregation. In particular, Aβ42 shows remarkably high affinity to copper- one molecule of Aβ42 was shown to bind 2.5 copper ions on an average [89]. Metal chelators such as EDTA [90], desferrioxamine [91], and clioquinol [92] were shown to reverse Aβ aggregation. In addition, metal chelators inhibit the interaction between Aβ and the lipid membrane [93], which is believed to catalyze Aβ aggregation, and hence offer a promising treatment for reducing the clinical symptoms of AD. Despite their promise in treating AD, metal chelators have low CNS bioavailability and pose significant systemic toxicity. D-penicillamine, a copper chelator, covalently conjugated to 1,2-Dioleoyl-sn-glycero-3-phosphoethanolamine-N-[4-(p-maleimidophenyl)butyramide] or 1,2-dioleoyl-sn-glycero-3-phosphoethanolamine-N-[3-(2 pyridyldithio)propionate] nanoparticles *via* a disulfide bond or a thioether bond demonstrated the ability to disaggregate Aβ plaques *in vitro* [94]. The nanoparticle surface serves as a backbone to bi-dentate iron chelators such as 2-methyl-N-(2'-aminoethyl or 3'-aminopropyl)-3-hydroxyl-4-pyridinone and transforms them into hexandentate chelators [95, 96], which were shown to have improved metal binding and lower toxicity than the bi-dentate chelators [97].

### 6.2.2 β-Sheet Breakers

Tjernberg's peptide (KLVFF) and Soto's peptide (LPFFD), also known as β-sheet breaker peptides, were proposed to interact with the aromatic structures and salt bridges within Aβ proteins and inhibit the formation of fibrils. Despite their *in vitro* efficacy, these short peptides suffer from two major disadvantages *in vivo*: (a) rapid proteolytic degradation in plasma, and (b) poor BBB permeability [98]. A nanodelivery system capable of permeating the BBB may not only protect these peptides from plasma degradation but also be helpful in delivering them to the brain parenchyma.

### 6.3 Strategies to Increase Aβ Clearance

Methods to increase removal of Aβ proteins from the brain include priming the immune system to produce anti-amyloid antibodies (active immunization) or the exogenous administration of anti-amyloid antibodies (passive immunization). Active immunization is achieved when a polyclonal anti-Aβ antibody is generated in the body of an AD subject following the administration of a full-length Aβ protein or an Aβ fragment coupled with an adjuvant. Initial studies, conducted by Dale Schenk in 1999, demonstrated that pre-aggregated human Aβ42 injected into a transgenic AD mouse lowered the amyloid plaque burden [99]. Encouraged by the success in pre-clinical trials, an active vaccine, AN1742 containing Aβ42 was developed and successfully tested in Phase I clinical trials. However, in phase II clinical trials, nearly 5% of the patients experienced aseptic meningoencephalitis and the clinical trial was immediately stopped [100]. To reduce these severe immune reactions casued by full length Aβ proteins, fragment of Aβ protein attached to a carrier protein were developed as vaccines [101]. Nanoparticles can improve the delivery of active vaccines to the brain. Songjiang and Lixiang developed fluorescently labeled chitosan nanoparticles loaded with subfragments of Aβ, which are believed to provide protection against AD [102]. This nanoparticle system allowed for a fourfold increase in the brain uptake as compared to the naked Aβ fragment.

Even though, active immunization paradigm showed promising results, further investigation is needed to ascertain its safety and efficacy in a large cohort of patients. Alternatively, passive immunization with the humanized anti-amyloid antibodies produced in a vector animal may be safer, because their dose and pharmacokinetics can be carefully monitored and regulated. At present, humanized monoclonal antibody AAB-001 (bapineuzumab), designed to bind to and remove Aβ from the brain is currently in Phase III clinical trials [103]. The anti-amyloid antibodies are more effective in reducing the parenchymal plaques as well as cerebrovascular amyloid deposits when they are delivered to the brain [104]. Our studies have shown that the delivery of a novel anti-amyloid antibody (IgG4.1) to the brain increased significantly upon conjugating with a naturally occurring polyamine such as putrescine. Alternatively, IgG4.1 could also be delivered to the brain *via* the smart nanovehicle described elsewhere in the article. Since these chitosan-based nanovehicles could be prepared under exceptionally mild conditions, the integrity of the antibodies is well preserved [59].

## 7. NATURAL REMEDIES

Various polyphenolic compounds are believed to offer protection against oxidative stress and neurotoxicity caused by Aβ proteins [105]. The beneficial effects of curcumin, resveratrol, and green tea extract against AD has been established with strong scientific evidence. Curcumin is a lipophilic polyphenol obtained from the roots of curcuma longa. Cole and associates have conducted extensive studies to demonstrate the efficacy of curcumin in reducing the formation of Aβ aggregates in AD transgenic mice [106]. Moreover, chronic curcumin administration was shown to lower oxidative stress and amyloid deposition in 16-month-old AD transgenic mice [107]. Although efficacious in reducing AD pathology, curcumin has low oral bioavailability because it undergoes extensive first pass metabolism. Encapsulation of curcumin in nanoparticles was shown to substantially enhance oral bioavailability and potentially improve the therapeutic efficacy of curcumin [108].

Resveratrol (trans-3,5,4-trihydroxystilbene) exerts anti-oxidant, anti-inflammatory, and anti-carcinogenic properties. Several studies have documented the efficacy of resveratrol in reducing AD pathology (reviewed by vingtdeux *et al.* 2008 [109]). *In vitro* studies conducted on rat hippocampal neurons demonstrated the efficacy of resveratrol in reducing Aβ induced neurotoxicity [110]. Kim *et al.* have reported that an intracerebral injection of resveratrol reduced hippocampal neurodegeneration in AD transgenic animals [111]. However, the major hurdle to realizing these therapeutic benefits of resveratrol is its low bioavailability. Like curcumin, the bioavailability of resveratrol could be significantly enhanced when formulated in nanoparticles [112], which may be critical for its use in AD therapy.

## CONCLUSIONS

The treatment and diagnosis of AD have seen significant improvements in the recent times. Nanotechnology is believed to hold promise to further advances in this field by providing opportunities for the early diagnosis and treatment of AD and CAA. Studies have demonstrated that nanotechlogy:

a) Contributes to the design of nanoprobes that can target histopathological hallmarks such as amyloid plaques, which are formed much before the appearance of neurological deficits, and aid in their detection *via* MRI and PET with high sensitivity and specificity;

b) Aids in the detection of various AD biomarkers in physiological fluids, which when combined with MRI and PET modalities could provide an accurate diagnosis of AD;

c) Helps the design of nanoparticles that serve as transport vehicles for a variety of therapeutic agents across the BBB and deliver them to the amyloid ridden tissue. By doing so, these nanoparticles are expected increase the efficacy of AD drugs while reducing their systemic toxicity.

Further gains in understanding AD pathophysiology coupled with future advances in nanotechnology are expected to turn AD into a detectable and treatable disorder.

## ACKNOWLEDGEMENTS

We sincerely acknowledge Dr. Joseph F. Poduslo and his labmates at the department of Neurology, Mayo Clinic, Rochester MN, USA for providing the images of MIONS.

## ABBREVIATIONS

**ABCA1** = ATP Binding Cassette Transporter 1, **AD** = Alzheimer's disease, **ADDLs** = Amyloid beta Derived Diffusible Ligands, **ApoE** = Apolipoprotein E, **APP** = Amyloid Precurssor Protein; **ATP** = Adenosine Triphosphate; **Aβ40** = Beta Amyloid 1-40; **Aβ** = Beta Amyloid protein; **Aβ42** = Beta Amyloid 1-42; **AβOs** = oligomers composed of 15–20 monomers; **BBB** = Blood Brain Barrier; **CAA** = Cerebral

Amyloid Angiopathy; **CED** = Convection Enhanced Delivery; **CNS** = Central Nervous System; **CPC** = Chitosan Polymeric Core; **CSF** = Cerebrospinal Fluid; **Da** = Daltons; **DFO** = Desferrioxamine; **DNA** = Deoxy Ribonucleic Acid; **ELISA** = Enzyme Linked Immunosorbent Assay; **F(ab')₂** = Fragment Antigen Binding 2; **Gd-DTPA** = Gadolinium diethylenetriaminepentaacetic acid; **HC** = Healthy Controls; **HIV** = Human Immunovirus; **IDE** = Insulin Degrading Enzyme; **IgG4.1** = Immunoglobulin G 4.1 antibody; **JAM** = Junctional Adhesion Molecules; **KDa** = Kilo Daltons; **LRP** = Low density Lipoprotein Related Protein Receptor; **LSPR** = Laser Surface Plasmon Resonance Spectroscopy; **LXR** = Liver X Receptors; **MIONS** = Monocrystalline Iron Oxide Nanoparticles; **MR** = Magnetic Resonance; **MRI** = Magnetic Resonance Imaging; **NEP** = Neprilysin; **NFT** = Neurofibrillary Tangles; **NIR** = Near Infrared; **PEG** = Polyethylene Glycol; **PET** = Positron Emisstion Tomography; **PIB** = Pittsburgh Compound B; **PLGA** = Polylactic-co-glycolic acid; **PSEN1** = Presenellin 1; **PSEN2** = Presenellin 2; **QDots** = Quantum Dots; **RAGE** = Receptor for Advanced Glycosylation Endproducts; **sAβ** = Soluble Beta Amyloid Protein; **SNV** = Smart Nanovehicles; **TfR** = Transferrin Receptor; **α2M** = Alpha-2-Macroglobulin; **€4** = allelle4; Nanorobots for Endovascular Target Interventions in Future Medical Practice

## REFERENCES

[1]    Graeber MB, Kosel S, Egensperger R, *et al.* Rediscovery of the case described by Alois Alzheimer in 1911: historical, histological and molecular genetic analysis. Neurogenetics 1997;1(1):73-80.

[2]    Brookmeyer R, Johnson E, Ziegler-Graham K, *et al.* Forecasting the global burden of Alzheimer's disease. Alzheimers Dement 2007;3(3):186-91.

[3]    Evans DA, Funkenstein HH, Albert MS, *et al.* Prevalence of Alzheimer's disease in a community population of older persons. Higher than previously reported. JAMA 1989;262(18):2551-6.

[4]    McKhann G, Drachman D, Folstein M, *et al.* Clinical diagnosis of Alzheimer's disease: report of the NINCDS-ADRDA Work Group under the auspices of Department of Health and Human Services Task Force on Alzheimer's Disease. Neurology 1984;34(7):939-44.

[5]    Shepherd C, McCann H, Halliday GM. Variations in the neuropathology of familial Alzheimer's disease. Acta Neuropathol 2009;118(1):37-52.

[6]    Bird TD. Genetic aspects of Alzheimer disease. Genet Med 2008;10(4):231-9.

[7]    Koedam EL, Lauffer V, van der Vlies AE, *et al.* Early-versus late-ons, *et al*zheimer's disease: more than age alone. J Alzheimers Dis 2010;19(4):1401-8.

[8]    Ignatius MJ, Gebicke-Harter PJ, Skene JH, *et al.* Expression of apolipoprotein E during nerve degeneration and regeneration. Proc Natl Acad Sci USA 1986;83(4):1125-9.

[9]    Riddell DR, Zhou H, Atchison K, *et al.* Impact of apolipoprotein E (ApoE) polymorphism on brain ApoE levels. J Neurosci 2008;28(45):11445-53.

[10]   Fukumoto H, Ingelsson M, Garevik N, *et al.* APOE epsilon 3/ epsilon 4 heterozygotes have an elevated proportion of apolipoprotein E4 in cerebrospinal fluid relative to plasma, independent of Alzheimer's disease diagnosis. Exp Neurol 2003;183(1):249-53.

[11]   Saunders AM, Strittmatter WJ, Schmechel D, *et al.* Association of apolipoprotein E allele epsilon 4 with late-onset familial and sporadic Alzheimer's disease. Neurology 1993;43(8):1467-72.

[12]   Khachaturian AS, Corcoran CD, Mayer LS, *et al.* Apolipoprotein E epsilon4 count affects age at onset of Alzheimer disease, but not lifetime susceptibility: the Cache County Study. Arch Gen Psychiatry 2004;61(5):518-24.

[13]   Rossor MN, Revesz T, Lantos PL, *et al.* Semantic dementia with ubiquitin-positive tau-negative inclusion bodies. Brain 2000 Feb;123 ( Pt 2):267-76.

[14]   Prada CM, Garcia-Alloza M, Betensky RA, , *et al.* Antibody-mediated clearance of amyloid-beta peptide from cerebral amyloid angiopathy revealed by quantitative *in vivo* imaging. J Neurosci 2007 Feb 21;27(8):1973-80.

[15]   Revesz T, Ghiso J, Lashley T , *et al.* Cerebral amyloid angiopathies: a pathologic, biochemical, and genetic view. J Neuropathol Exp Neurol 2003;62(9):885-98.

[16]   Thal DR, Griffin WS, de Vos RA , *et al.* Cerebral amyloid angiopathy and its relationship to Alzheimer's disease. Acta Neuropathol 2008;115(6):599-609.

[17]   Haass C, Selkoe DJ. Cellular processing of beta-amyloid precursor protein and the genesis of amyloid beta-peptide. Cell 1993;75(6):1039-42.

[18]   Gouras GK, Xu H, Jovanovic JN , *et al.* Generation and regulation of beta-amyloid peptide variants by neurons. J Neurochem. 1998;71(5):1920-5.

[19]   Glabe CC. Amyloid accumulation and pathogensis of Alzheimer's disease: significance of monomeric, oligomeric and fibrillar Abeta. Subcell Biochem 2005;38:167-77.

[20]   Walsh DM, Selkoe DJ. A beta oligomers - a decade of discovery. J Neurochem 2007;101(5):1172-84.

[21]   Lambert MP, Barlow AK, Chromy BA, *et al*. Diffusible, nonfibrillar ligands derived from Abeta1-42 are potent central nervous system neurotoxins. Proc Natl Acad Sci USA 1998;95(11):6448-53.

[22]   Kayed R, Head E, Thompson JL, *et al*. Common structure of soluble amyloid oligomers implies common mechanism of pathogenesis. Science 2003 Apr 18;300(5618):486-9.

[23]   Nguyen HD, Hall CK. Kinetics of fibril formation by polyalanine peptides. J Biol Chem 2005;280(10):9074-82.

[24]   Lesne S, Koh MT, Kotilinek L, *et al*. A specific amyloid-beta protein assembly in the brain impairs memory. Nature 2006;440(7082):352-7.

[25]   Hartley DM, Walsh DM, Ye CP, *et al*. Protofibrillar intermediates of amyloid beta-protein induce acute electrophysiological changes and progressive neurotoxicity in cortical neurons. J Neurosci 1999;19(20):8876-84.

[26]   Wirths O, Multhaup G, Bayer TA. A modified beta-amyloid hypothesis: intraneuronal accumulation of the beta-amyloid peptide--the first step of a fatal cascade. J Neurochem 2004;91(3):513-20.

[27]   LaFerla FM, Green KN, Oddo S. Intracellular amyloid-beta in Alzheimer's disease. Nat Rev Neurosci 2007;8(7):499-509.

[28]   LaFerla FM, Troncoso JC, Strickland DK, *et al*. Neuronal cell death in Alzheimer's disease correlates with apoE uptake and intracellular Abeta stabilization. J Clin Invest 1997;100(2):310-20.

[29]   Guo Q, Fu W, Xie J, *et al*. Par-4 is a mediator of neuronal degeneration associated with the pathogenesis of Alzheimer disease. Nat Med 1998;4(8):957-62.

[30]   Herzig MC, Winkler DT, Burgermeister P , *et al*. Abeta is targeted to the vasculature in a mouse model of hereditary cerebral hemorrhage with amyloidosis. Nat Neurosci 2004;7(9):954-60.

[31]   Tanzi RE, Moir RD, Wagner SL. Clearance of Alzheimer's Abeta peptide: the many roads to perdition. Neuron 2004;43(5):605-8.

[32]   Giedraitis V, Sundelof J, Irizarry MC, *et al*. The normal equilibrium between CSF and plasma amyloid beta levels is disrupted in Alzheimer's disease. Neurosci Lett 2007 Nov 12;427(3):127-31.

[33]   DeMattos RB. Apolipoprotein E dose-dependent modulation of beta-amyloid deposition in a transgenic mouse model of Alzheimer's disease. J Mol Neurosci. 2004;23(3):255-62.

[34]   Deane R, Du Yan S, Submamaryan RK, *et al*. RAGE mediates amyloid-beta peptide transport across the blood-brain barrier and accumulation in brain. Nat Med 2003;9(7):907-13.

[35]   Deane R, Bell RD, Sagare A, *et al*. Clearance of amyloid-beta peptide across the blood-brain barrier: implication for therapies in Alzheimer's disease. CNS Neurol Disord Drug Targets 2009;8(1):16-30.

[36]   Faller P. Copper and zinc binding to amyloid-beta: coordination, dynamics, aggregation, reactivity and metal-ion transfer. Chembiochem 2009;10(18):2837-45.

[37]   Jun S, Gillespie JR, Shin BK, *et al*. The second Cu(II)-binding site in a proton-rich environment interferes with the aggregation of amyloid-beta(1-40) into amyloid fibrils. Biochemistry 2009;48(45):10724-32.

[38]   D'Andrea MR, Nagele RG, Wang HY, *et al*. Evidence that neurones accumulating amyloid can undergo lysis to form amyloid plaques in Alzheimer's disease. Histopathology 2001;38(2):120-34.

[39]   Kantarci K, Jack CR, Jr. Quantitative magnetic resonance techniques as surrogate markers of Alzheimer's disease. NeuroRx 2004;1(2):196-205.

[40]   Jack CR, Jr., Wengenack TM, Reyes DA, *et al*. *In vivo* magnetic resonance microimaging of individual amyloid plaques in Alzheimer's transgenic mice. J Neurosci 2005;25(43):10041-8.

[41]   Poduslo JF, Wengenack TM, Curran GL, *et al*. Molecular targeting of Alzheimer's amyloid plaques for contrast-enhanced magnetic resonance imaging. Neurobiol Dis 2002;11(2):315-29.

[42]   Wengenack TM, Curran GL, Poduslo JF. Targeting alzheimer amyloid plaques *in vivo*. Nat Biotechnol 2000;18(8):868-72.

[43]   Poduslo JF, Curran GL, Peterson JA, *et al*. Design and chemical synthesis of a magnetic resonance contrast agent with enhanced *in vitro* binding, high blood-brain barrier permeability, and *in vivo* targeting to Alzheimer's disease amyloid plaques. Biochemistry 2004;43(20):6064-75.

[44]   Lee HJ, Zhang Y, Zhu C, *et al*. Imaging brain amyloid of alzheimer disease *in vivo* in transgenic mice with an A[bgr] peptide radiopharmaceutical. J Cereb Blood Flow Metab 2002;22(2):223-31.

[45]   Wadghiri YZ, Sigurdsson EM, Sadowski M, *et al*. Detection of Alzheimer's amyloid in transgenic mice using magnetic resonance microimaging. Magn Reson Med 2003;50(2):293-302.

[46]   Lockhart A, Lamb JR, Osredkar T, *et al*. PIB is a non-specific imaging marker of amyloid-beta (Abeta) peptide-related cerebral amyloidosis. Brain 2007;130(Pt 10):2607-15.

[47] Price JC, Klunk WE, Lopresti BJ, *et al*. Kinetic modeling of amyloid binding in humans using PET imaging and Pittsburgh Compound-B. J Cereb Blood Flow Metab 2005;25(11):1528-47.

[48] Grimmer T, Henriksen G, Wester HJ, *et al*. Clinical severity of Alzheimer's disease is associated with PIB uptake in PET. Neurobiol Aging 2009;30(12):1902-9.

[49] Engler H, Forsberg A, Almkvist O, *et al*. Two-year follow-up of amyloid deposition in patients with Alzheimer's disease. Brain. 2006;129(Pt 11):2856-66.

[50] Klunk WE, Engler H, Nordberg A, *et al*. Imaging brain amyloid in Alzheimer's disease with Pittsburgh Compound-B. Ann Neurol 2004;55(3):306-19.

[51] Bareford LM, Swaan PW. Endocytic mechanisms for targeted drug delivery. Adv Drug Deliv Rev 2007;59(8):748-58.

[52] Aktas Y, Yemisci M, Andrieux K, *et al*. Development and brain delivery of chitosan-PEG nanoparticles functionalized with the monoclonal antibody OX26. Bioconjug Chem 2005;16(6):1503-11.

[53] Pang Z, Lu W, Gao H , *et al*. Preparation and brain delivery property of biodegradable polymersomes conjugated with OX26. J Control Release 2008;128(2):120-7.

[54] Skaat H, Margel S. Synthesis of fluorescent-maghemite nanoparticles as multimodal imaging agents for amyloid-beta fibrils detection and removal by a magnetic field. Biochem Biophys Res Commun 2009;386(4):645-9.

[55] Shubayev VI, Pisanic TR, 2nd, Jin S. Magnetic nanoparticles for theragnostics. Adv Drug Deliv Rev 2009;61(6):467-77.

[56] Gupta AK, Gupta M. Synthesis and surface engineering of iron oxide nanoparticles for biomedical applications. Biomaterials 2005;26(18):3995-4021.

[57] Molday RS, MacKenzie D. Immunospecific ferromagnetic iron-dextran reagents for the labeling and magnetic separation of cells. J Immunol Methods 1982;52(3):353-67.

[58] Chan WC, Maxwell DJ, Gao X, *et al*. Luminescent quantum dots for multiplexed biological detection and imaging. Curr Opin Biotechnol 2002;13(1):40-6.

[59] Gao X, Chan WC, Nie S. Quantum-dot nanocrystals for ultrasensitive biological labeling and multicolor optical encoding. J Biomed Opt 2002;7(4):532-7.

[60] Hezinger AF, Tessmar J, Gopferich A. Polymer coating of quantum dots--a powerful tool toward diagnostics and sensorics. Eur J Pharm Biopharm 2008;68(1):138-52.

[61] Agyare EK, Curran GL, Ramakrishnan M, *et al*. Development of a smart nano-vehicle to target cerebrovascular amyloid deposits and brain parenchymal plaques observed in Alzheimer's disease and cerebral amyloid angiopathy. Pharm Res 2008;25(11):2674-84.

[62] Verbeek MM, Olde Rikkert MG. Cerebrospinal fluid biomarkers in the evaluation of Alzheimer disease. Clin Chem. 2008;54(10):1589-91.

[63] Parnetti L, Lanari A, Amici S, *et al*. CSF phosphorylated tau is a possible marker for discriminating Alzheimer's disease from dementia with Lewy bodies. Phospho-Tau International Study Group. Neurol Sci 2001;22(1):77-8.

[64] Haes AJ, Chang L, Klein WL, *et al*. Detection of a biomarker for Alzheimer's disease from synthetic and clinical samples using a nanoscale optical biosensor. J Am Chem Soc 2005;127(7):2264-71.

[65] Nam JM, Thaxton CS, Mirkin CA. Nanoparticle-based bio-bar codes for the ultrasensitive detection of proteins. Science 2003;301(5641):1884-6.

[66] Georganopoulou DG, Chang L, Nam JM , *et al*. Nanoparticle-based detection in cerebral spinal fluid of a soluble pathogenic biomarker for Alzheimer's disease. Proc Natl Acad Sci USA 2005;102(7):2273-6.

[67] Faraji AH, Wipf P. Nanoparticles in cellular drug delivery. Bioorg Med Chem 2009;17(8):2950-62.

[68] Cuajungco MP, Faget KY, Huang X, *et al*. Metal chelation as a potential therapy for Alzheimer's disease. Ann N Y Acad Sci 2000;920:292-304.

[69] Drummond DC, Meyer O, Hong K, *et al*. Optimizing liposomes for delivery of chemotherapeutic agents to solid tumors. Pharmacol Rev 1999;51(4):691-743.

[70] Harush-Frenkel O, Rozentur E, Benita S, *et al*. Surface charge of nanoparticles determines their endocytic and transcytotic pathway in polarized MDCK cells. Biomacromolecules 2008;9(2):435-43.

[71] Alexis F, Pridgen E, Molnar LK, *et al*. Factors affecting the clearance and biodistribution of polymeric nanoparticles. Mol Pharm 2008;5(4):505-15.

[72] Dehouck B, Fenart L, Dehouck MP, *et al*. A new function for the LDL receptor: transcytosis of LDL across the blood-brain barrier. J Cell Biol 1997;138(4):877-89.

[73] Duffy KR, Pardridge WM. Blood-brain barrier transcytosis of insulin in developing rabbits. Brain Res 1987;420(1):32-8.

[74]    Frank HJ, Pardridge WM, Morris WL, *et al.* Binding and internalization of insulin and insulin-like growth factors by isolated brain microvessels. Diabetes 1986;35(6):654-61.

[75]    Ulbrich K, Hekmatara T, Herbert E, *et al.* Transferrin- and transferrin-receptor-antibody-modified nanoparticles enable drug delivery across the blood-brain barrier (BBB). Eur J Pharm Biopharm 2009;71(2):251-6.

[76]    Omori N, Maruyama K, Jin G, *et al.* Targeting of post-ischemic cerebral endothelium in rat by liposomes bearing polyethylene glycol-coupled transferrin. Neurol Res 2003;25(3):275-9.

[77]    Zhang Y, Zhang YF, Bryant J, *et al.* Intravenous RNA interference gene therapy targeting the human epidermal growth factor receptor prolongs survival in intracranial brain cancer. Clin Cancer Res 2004;10(11):3667-77.

[78]    Wilson B, Samanta MK, Santhi K, *et al.* Poly(n-butylcyanoacrylate) nanoparticles coated with polysorbate 80 for the targeted delivery of rivastigmine into the brain to treat Alzheimer's disease. Brain Res 2008;1200:159-68.

[79]    Kreuter J. Nanoparticulate systems for brain delivery of drugs. Adv Drug Deliv Rev 2001;47(1):65-81.

[80]    Wilson B, Samanta MK, Santhi K, Kumar KP, Ramasamy M, Suresh B. Chitosan nanoparticles as a new delivery system for the anti-Alzheimer drug tacrine. Nanomedicine 2010;6(1):144-52.

[81]    Kim D, Tsai LH. Bridging physiology and pathology in AD. Cell 2009;137(6):997-1000.

[82]    Xiong M, Zhang T, Zhang LM, *et al.* Caspase inhibition attenuates accumulation of beta-amyloid by reducing beta-secretase production and activity in rat brains after stroke. Neurobiol Dis 2008;32(3):433-41.

[83]    Anderson JJ, Holtz G, Baskin PP, *et al.* Reductions in beta-amyloid concentrations *in vivo* by the gamma-secretase inhibitors BMS-289948 and BMS-299897. Biochem Pharmacol 2005;69(4):689-98.

[84]    Eli Lilly and Company assignee. Effect of γ-Secretase Inhibition on the Progression of Alzheimer's Disease: LY450139 Versus Placebo 2008.

[85]    Tanabe C, Ebina M, Asai M, *et al.* 1,3-Capryloyl-2-arachidonoyl glycerol activates alpha-secretase activity and suppresses Abeta40 secretion in A172 cells. Neurosci Lett 2009;450(3):324-6.

[86.    Miller VM, Gouvion CM, Davidson BL, *et al.* Targeting Alzheimer's disease genes with RNA interference: an efficient strategy for silencing mutant alleles. Nucleic Acids Res 2004;32(2):661-8.

[87]    Akhtar S, Benter IF. Nonviral delivery of synthetic siRNAs *in vivo*. J Clin Invest 2007;117(12):3623-32.

[88]    Prakash S, Malhotra M, Rengaswamy V. Nonviral siRNA delivery for gene silencing in neurodegenerative diseases. Methods Mol Biol 2010;623:211-29.

[89]    Singer O, Marr RA, Rockenstein E, *et al.* Targeting BACE1 with siRNAs ameliorates Alzheimer disease neuropathology in a transgenic model. Nat Neurosci 2005;8(10):1343-9.

[90]    Chauhan NB, Siegel GJ. Antisense inhibition at the beta-secretase-site of beta-amyloid precursor protein reduces cerebral amyloid and acetyl cholinesterase activity in Tg2576. Neuroscience 2007;146(1):143-51.

[91]    Atwood CS, Scarpa RC, Huang X, *et al.* Characterization of copper interactions with alzheimer amyloid beta peptides: identification of an attomolar-affinity copper binding site on amyloid beta1-42. J Neurochem 2000;75(3):1219-33.

[92]    Casdorph HR. EDTA chelation therapy: efficacy in brain disorders. In: EM C, Ed. A Textbook on EDTA Chelation Therapy. Charlottesville: Hamtpon Roads Publishing Company, Inc. 2001; pp. 142–63.

[93]    Mandel S, Amit T, Bar-Am O, *et al.* Iron dysregulation in Alzheimer's disease: multimodal brain permeable iron chelating drugs, possessing neuroprotective-neurorescue and amyloid precursor protein-processing regulatory activities as therapeutic agents. Prog Neurobiol 2007;82(6):348-60.

[94]    Cherny RA, Atwood CS, Xilinas ME, *et al.* Treatment with a copper-zinc chelator markedly and rapidly inhibits beta-amyloid accumulation in Alzheimer's disease transgenic mice. Neuron 2001;30(3):665-76.

[95]    Deraeve C, Pitie M, Meunier B. Influence of chelators and iron ions on the production and degradation of $H_2O_2$ by beta-amyloid-copper complexes. J Inorg Biochem 2006;100(12):2117-26.

[96]    Cui Z, Lockman PR, Atwood CS, *et al.* Novel D-penicillamine carrying nanoparticles for metal chelation therapy in Alzheimer's and other CNS diseases. Eur J Pharm Biopharm 2005;59(2):263-72.

[97]    Liu G, Men P, Perry G, *et al.* Metal chelators coupled with nanoparticles as potential therapeutic agents for Alzheimer's disease. J Nanoneurosci 2009;1(1):42-55.

[98]    Liu G, Men P, Kudo W, *et al.* Nanoparticle-chelator conjugates as inhibitors of amyloid-beta aggregation and neurotoxicity: a novel therapeutic approach for Alzheimer disease. Neurosci Lett 2009;455(3):187-90.

[99]    Hider RC, Epemolu O, Singh S, *et al* JB. Iron chelator design. Adv Exp Med Biol 1994;356:343-9.

[100]    Poduslo JF, Curran GL, Kumar A, *et al.* Beta-sheet breaker peptide inhibitor of Alzheimer's amyloidogenesis with increased blood-brain barrier permeability and resistance to proteolytic degradation in plasma. J Neurobiol 1999;39(3):371-82.

[101]    Schenk D, Barbour R, Dunn W, *et al.* Immunization with amyloid-beta attenuates Alzheimer-disease-like pathology in the PDAPP mouse. Nature 1999;400(6740):173-7.

[102] Steinberg D. Companies halt first Alzheimer vaccine trial. Scientist 2002;16:22.

[103] Petrushina I, Ghochikyan A, Mkrtichyan M, *et al.* Mannan-Abeta28 conjugate prevents Abeta-plaque deposition, but increases microhemorrhages in the brains of vaccinated Tg2576 (APPsw) mice. J Neuroinflammation 2008;5:42.

[104] Songjiang Z, Lixiang W. Amyloid-beta associated with chitosan nano-carrier has favorable immunogenicity and permeates the BBB. AAPS PharmSciTech 2009;10(3):900-5.

[105] JANSSEN Alzheimer Immunotherapy Research & Development, LLC assignee. A Phase 3, Multicenter, Randomized, Double-Blind, Placebo-Controlled, Parallel-Group, Efficacy and Safety Trial of Bapineuzumab (AAB-001, ELN115727) In Patients With Mild to Moderate Alzheimer's Disease Who Are Apolipoprotein E4 Non- Carriers 2007.

[106] Thakker DR, Weatherspoon MR, Harrison J, *et al.* Intracerebroventricular amyloid-beta antibodies reduce cerebral amyloid angiopathy and associated micro-hemorrhages in aged Tg2576 mice. Proc Natl Acad Sci USA 2009;106(11):4501-6.

[107] Butterfield D, Castegna A, Pocernich C , *et al.* Nutritional approaches to combat oxidative stress in Alzheimer's disease. J Nutr Biochem 2002;13(8):444.

[108] Yang F, Lim GP, Begum AN, *et al.* Curcumin inhibits formation of amyloid beta oligomers and fibrils, binds plaques, and reduces amyloid *in vivo.* J Biol Chem 2005;280(7):5892-901.

[109] Lim GP, Chu T, Yang F, *et al.* The curry spice curcumin reduces oxidative damage and amyloid pathology in an Alzheimer transgenic mouse. J Neurosci 2001;21(21):8370-7.

[110] Shaikh J, Ankola DD, Beniwal V, *et al.* Nanoparticle encapsulation improves oral bioavailability of curcumin by at least 9-fold when compared to curcumin administered with piperine as absorption enhancer. Eur J Pharm Sci 2009;37(3-4):223-30.

[111] Vingtdeux V, Dreses-Werringloer U, Zhao H, *et al.* Therapeutic potential of resveratrol in Alzheimer's disease. BMC Neurosci 2008;9 (Suppl 2):S6.

[112] Han YS, Zheng WH, Bastianetto S, *et al.* Neuroprotective effects of resveratrol against beta-amyloid-induced neurotoxicity in rat hippocampal neurons: involvement of protein kinase C. Br J Pharmacol 2004;141(6):997-1005.

[113] Kim D, Nguyen MD, Dobbin MM, *et al.* SIRT1 deacetylase protects against neurodegeneration in models for Alzheimer's disease and amyotrophic lateral sclerosis. EMBO J 2007;26(13):3169-79.

[114] Teskac K, Kristl J. The evidence for solid lipid nanoparticles mediated cell uptake of resveratrol. Int J Pharm 2010;390(1):61-9.

# Index

**A**

A-beta proteins 109
Alzheimer's disease 107
Angiogenesis 6
Antimicrobial activity 28
Atherosclerosis 3

**B**

Beta sheet breakers 118
Biomarkers 114
Blood-tumour barrier 43

**C**

Cancer cells 11
Cardiology 34
Cardiovascular 6
Chemotherapy 10
Coronary in-stent restenosis 32
Cytotoxicity 75

**D**

Dendrimer nanoparticles 11
Doxorubicin 11,

**E**

EGF receptor
Endocytosis 77
Endothelial cells 74

**F**

Ferromagnetic bead 95

**G**

Gadolinium nanoparticles 52, 64
Gold nanoparticles 48

**H**

Her2 54

**I**

Imaging amyloid plaques 111
Inflammation 35
Iron oxide nanocrystals 43, 53

**L**

Liposomes 11